Agricultural
Research
Alternatives

William Lockeretz and
Molly D. Anderson

Agricultural
Research
Alternatives

University of Nebraska Press

Lincoln and London

Library of Congress Cataloging-in-Publication Data
Lockeretz, William.
Agricultural research alternatives / William Lockeretz
and Molly D. Anderson.
p.   cm.—(Our sustainable future: v. 3)
Includes bibliographical references and index.
ISBN 0-8032-2901-1 (cl: alk. paper)
1. Agriculture—Research.   I. Anderson, Molly D.
(Molly DelCarmen), 1955–   .   II. Title.   III. Series.
S540.A2L63   1993
630'.72—dc20
92-47113   CIP

• • •

In memory of Frank Pipkin (1925–1992):

an exemplary scientist, a sympathetic adviser,

and a kindly human being

Contents

. . .

Acknowledgments

We are grateful for the generous financial support from three foundations that made it possible for us to write this book. Initial funding came from the Northwest Area Foundation; support to complete the project was provided by the Ford Foundation and the Joyce Foundation. A particularly important role was played by Karl Stauber, of the Northwest Area Foundation, who saw the desirability of a long-range look at approaches to agricultural research and let us take the time we thought was needed.

Many people at universities, research institutions, and advocacy organizations reviewed draft chapters, offering a wide range of helpful comments. Even when we chose not to follow a suggestion, it was very valuable to rethink the point after having heard a differing view. To all these people, our sincere thanks: Jill Auburn, Jim Bender, Norman Berg, Elizabeth Bird, James Bonnen, Harold Breimyer, Robert Bugg, Jeffrey Burkhardt, Frances Chew, Katherine Clancy, Kent Crookston, David Danbom, Thomas Dobbs, Otto Doering, Sara Ebenreck, Franklin Eggert, Charles Francis, John Gardner, John Gerber, Gilbert Gillespie, Richard Harwood, Peter Hildebrand, Stephen Kaffka, Dennis Keeney, Larry King, Peter Korsching, Sheldon Krimsky, Harry Kunkel, William Liebhardt, Matt Liebman, Rod MacRae, Patrick Madden, Nanette Mengel, Robert Miller, William Murphy, Robert Papendick, James Parr, Eldor Paul, John Pesek, Beatrice Rogers, Vernon Ruttan, Richard Sauer, Neill Schaller, Wesley Seitz, Georgia Shearer, Karl

Stauber, Benjamin Stinner, Paul Thompson, Richard Thompson, David Vail, and Garth Youngberg. We are especially indebted to Sarah Wernick for the many stylistic and organizational changes she suggested throughout the manuscript.

Portions of several chapters have been adapted from previously published articles (see reference list for full citations): chapter 5, Lockeretz (1991b); chapter 7, Lockeretz (1991a); chapter 8, Anderson (1992) and Anderson and Lockeretz (1991); chapter 9, Lockeretz and Anderson (1990). This material is used with the permission of AB Academic Publishers, publisher of *Biological Agriculture and Horticulture,* and the Henry A. Wallace Institute for Alternative Agriculture, publisher of *American Journal of Alternative Agriculture.*

Part One

APPROACHES TO AGRICULTURAL
RESEARCH, PAST AND PRESENT

1

. . .

What May We Expect from
Agricultural Research?

There is reason to hope that we may presently reach a system of cultivation in which, though the crops may be large, the land itself shall not only not be exhausted, but be in a course of continual amelioration . . . without any extraneous aid, and from the resources of the farm itself.—Colman, 1856, pp. 222–23

For more than a century, the United States has strongly supported public agricultural research. This support reflects our belief that research is essential if agriculture is to meet society's ever-changing demands. Agriculture has been expected to adapt to all sorts of changes: cycles of good and bad economic times; the transition from a rural to an urban-industrial society; increasing competition for natural resources; rapid technological advances in the rest of the economy; and the expansion of global markets. Providing this adaptability has been the great mission of our public agricultural research institutions.

This task has become especially challenging in recent years because the expectations placed on agriculture have become more varied, pulling it in many directions at once. Agriculture still has its traditional job of producing an abundant food supply reliably and cheaply. Now, however, it is being asked to do many other things too: reduce its consumption of finite resources such as fossil fuel and water; avoid damage to the environment; minimize toxic residues in our food; reverse the deterioration of rural communities; and, more generally, preserve our long-term productive capacity.

3

Just as the expectations placed on research have become more varied, so too have the ways we judge its benefits. Traditionally, we have rated agriculture largely by short-term economic measures. What has counted most has been what agriculture produces and the costs of producing it, with the emphasis on the current year. Moreover, "costs" have meant mainly the things that farmers must pay for, not costs that are borne by others or that have no monetary value, such as environmental damage. In the past, what research contributed to national well-being often got condensed into a single number: crop yield per acre, or the number of people fed by each farmer, or the proportion of disposable income spent on food. Today, the criteria are not so exclusively internal to agriculture, and they are less dominated by economics. Also, they are less slanted toward the present at the possible expense of the future.

How well has public research improved agriculture's performance according to today's more diverse criteria? The question is difficult to answer. We know that agriculture's problems are far from solved. Soil erosion continues, along with contamination and depletion of groundwater, and harm to wildlife. Agriculture remains heavily dependent on fossil energy. The level of rural services continues to deteriorate, and the economy of many farm communities is chronically unsound, even though the acute distress that farmers suffered in the mid-1980s has eased.

Of course, it would be unfair to pin all these problems on agricultural research—there has been undeniable progress with some, and the solutions require more than research. But blaming research would not be entirely unfair. When things have gone well, agricultural researchers have not hesitated to claim credit. Often cited, for example, is the rapid increase in agricultural productivity and efficiency after World War II (mentioned by Jordan et al. [1986], among others). However, much of this improvement came from research and development outside agriculture, such as microelectronic sensors and controllers, cheaper methods for manufacturing fertilizer, and improved diesel engines. It therefore seems justified to make the inverse argument, too: the persistence of many problems can be regarded in part as a failure of agricultural research.

4

If researchers are to meet these unfulfilled needs, it will not be enough simply to give them more money—the solution commonly offered first by scientists in all fields. Nor do we simply need to draw up a new "research agenda" for the next few years, or to "redirect" existing funds. It is more a question of improving *how* the money is used.

Where might we look for innovative ideas to achieve this? For many people, the first answer is biotechnology, with its undisputed power to revolutionize agriculture. Its growing importance, however, does not diminish the role of more traditional research, which is what this book is about.

In the traditional domain, an encouraging development has been the progress in using on-farm resources to reduce the need for manufactured inputs, especially fertilizers and pesticides. Examples include pest management systems that allow farmers to omit some pesticide applications, legumes as an alternative to inorganic nitrogen fertilizer, and organic farming systems that avoid both synthetic pesticides and fertilizers.

Work on such systems represents a change in the subject matter, the "what" of research, but it also can contribute to the "how." The reason is that ideas about reducing chemical use often encompass broader social and environmental goals. We can see this in the overlapping terms "reduced-chemical" and "alternative" agriculture. The former is an objective characteristic of specific production methods; the latter implies more general changes in the entire agricultural system. Yet to most people, the term "alternative" agriculture first suggests reduced use of chemicals. Conversely, reduced use of chemicals by itself can qualify a system as "alternative."

Therefore, there may be more general value in the innovative thinking behind alternative and reduced-chemical techniques. As we explain in the next chapter, people who have been advocating more research on alternative systems also have ideas about how this research should be done. They favor a more multidisciplinary approach, greater attention to agroecological principles and farmers' use of information, and more research on-site in farmers' fields. Also, they often call on researchers to be more responsive and more accountable to groups outside the research establishment, especially farmers.

5

These recommendations are a provocative challenge to prevailing research approaches. But why restrict the new ideas to reduced-chemical or alternative systems? Why shouldn't we give greater attention to how farmers use information in any production system? Couldn't agroecology make an important contribution even if chemicals are used? Might not farmers' fields be good sites for many kinds of research? Perhaps the ideas offered by advocates of reduced-chemical alternatives can give new life to agricultural research as a whole. This book looks more closely at this intriguing possibility.

The reason for extending these ideas beyond alternative agriculture is that the process of research may not be as closely connected to its content as we often hear. In part, the connection may have arisen because the same people want to change both the process and the content. Initially, the call for more attention to alternative agriculture came largely from groups outside the research establishment, such as environmentalists, some farmers' organizations, safe food advocates, and rural activists. As outsiders, such people probably are more inclined to challenge accepted ideas about the professional and institutional side of research. Certainly, they would not agree that outside groups should have only a minor influence. So, too, they are not likely to sympathize when the profession rewards its members more for what it considers important than for how valuable the research is to society, particularly if this reward system steers researchers away from multidisciplinary research and other worthwhile but less-accepted approaches.

Still, even if the content and process of alternative agriculture research are linked in part because of circumstance, the two no doubt do have an underlying relationship. To reveal that relationship, we must strip away some exaggerations and oversimplifications that have become attached to alternative research approaches. Part 2 of this book shows that when we do so, the changes proposed for the research process claim both too much and too little. Too much, because strategies such as on-farm research are not the only way to study alternative agriculture. Too little, because they can be valuable far beyond research on alternative agriculture, provided they are applied discerningly, not as a cure-all.

But if we are to capture the benefit of these ideas, the institutional and

6

professional setting must be right. In part 3, therefore, we propose changes in departmental structure, professional rewards, grant programs, and education that could help research take advantage of innovative approaches to meet the many new demands being placed upon it.

A note of moderation is in order, however. When we suggest that meeting these demands will require some far-reaching alternatives to accepted ideas, it is not because every blip on agricultural trend charts means that the customary ways of doing research are outdated. The research approaches we discuss are intended to supplement current strategies, not replace them. A perspective of more than a few years is advisable before announcing that agricultural research needs a new "paradigm." The historical record, described in chapter 3, should temper the exuberance of people who believe that change is more rapid than ever before, thus demanding radical changes in research. Coping with the devastated economy of the Great Depression, for example, was at least as daunting as the challenges we face today. Still, the diversity of the current pressures on agriculture—economic and noneconomic, internal and external, short- and long-term—has little precedent, and argues for seriously rethinking the processes and institutional structures of agricultural research.

Is it realistic to expect such a reexamination? Agricultural research has not usually been given to self-scrutiny. Those who have criticized its emphasis on short-term technical and economic efficiency have largely been outside the research system. The system has typically responded with hostility and defensiveness or, at best, reluctant acceptance. In an encouraging development, however, challenges to the way we do agricultural research are now coming also from within the agricultural establishment. Many researchers and administrators have begun to think about fundamental, long-lasting changes that might revitalize agricultural research and prepare it to meet future demands. The question now is how to do the job well.

7

2

. . .

The Connection between Alternative Agricultural Systems and Alternative Research Approaches

God speed the time when we shall agree on some fundamental principles, and when we shall discover and demonstrate the best and most economic methods for the permanent maintenance or increase of the productive capacity of our soils . . . not by the purchase and use of sodium nitrate, almost certainly not, but undoubtedly by the assimilation and utilization of unlimited quantities of atmospheric nitrogen; probably not by the use of acid phosphates . . . but much more likely by returning to the land . . . farm manures and leguminous green fertilizers. . . . The American farmer has a right to expect that, if he adopts the methods which we advocate, the fertility of his soil is secure, that the productive capacity of his land will be increased, or, at the very least, that it shall be permanently maintained—not only for a season, not only for a score of years, but so long as the American farmer shall till American soil.
—Hopkins, 1904, pp. 103–4

This book is about how, where, and by whom agricultural research is done, not about particular production systems or research topics. However, its ideas for improving the research process have come from people interested in particular kinds of agricultural systems. In part 2, we loosen the connection, arguing that these ideas also apply to many other areas of research. But first we explore the origins of the connection to help clarify when and why particular research topics do call for particular research processes.

8

Alternative Agriculture and Related Concepts

The central ideas of this book come from what we will refer to as "alternative" agriculture, although we might have credited them instead to sustainable, low-input, ecological, or reduced-chemical agriculture. Much has been written about these labels. The same term may mean different things to different authors; conversely, one author may use several interchangeably (Lockeretz, 1988). They may designate the goals of an agricultural system, its production methods, or the values and attitudes of the farmers who use them. For some people, these terms refer mainly to the physical and biological consequences of an agricultural system, such as its effects on water quality and soil fertility. Others attach great importance to social justice, economic equity, and farm structure, often favoring small and medium-size farms that are family-owned and -operated (Crosson, 1989; Madden, 1989; Allen et al., 1991; Crews et al., 1991).

For our purposes, we need not get bogged down in definitions. We treat the various terms loosely, concentrating on the rich core of ideas held by most of their advocates. Their major goals include conservation of natural resources, protection of the environment, enhanced social and economic well-being of rural communities, and improved food quality. High on every list of how to achieve these goals is the reduction or elimination of pesticides and fertilizers; sometimes this strategy is generalized to a decreased use of all purchased inputs.

For simplicity, we label these ideas "alternative agriculture." We realize that the term is imprecise, but its basic meaning is understood well enough. We have chosen it over the others because it has an additional implication that is relevant here: an "alternative" is offered in opposition to something "conventional." Much of the impetus for alternative agriculture research came from outside the research establishment. Its supporters often see it not simply as another area of research but also as a challenge to entrenched attitudes and customary political and institutional relationships. This connotation of "alternative" will come up in several subsequent chapters.

9

Origins of Alternative Agriculture and Its Sources of Support

ORGANIC FARMING

Alternative agriculture is an extension—some would say a dilution—of organic farming. For some early advocates, organic farming implied unconventional and at times mystical ideas about relationships among people, farming, nature, and the cosmos. No doubt that is why organic farming did not make headway in mainstream thinking for several decades.

However, by the 1960s some farmers had adopted organic methods even though they did not share the philosophy. They had more pragmatic reasons, especially the harmful health effects of agrichemicals (Wernick and Lockeretz, 1977). Since the 1970s, the environmental benefits and economic success that some of these farmers achieve have received increasing publicity. This has made organic farming influential beyond the small number of farmers who practice it.

From organic farming, alternative agriculture derived its most characteristic feature: reduced use, if not elimination, of synthetic pesticides and fertilizers. Organic farming also contributed the idea that farm-generated resources should replace purchased inputs in general, not just chemicals, and it stressed that farms should be diversified. Diversification is associated with the design and management of a farm as an integrated system rather than as an assemblage of components (Scofield, 1986).

ACTIVIST MOVEMENTS

A second force behind alternative agriculture has been support from food safety advocates, environmentalists, and agrarian-oriented rural community groups and farmers' organizations. Food safety groups look to reduced-chemical alternatives to decrease the levels of pesticide residues, nitrates, and antibiotics in foods (Clancy, 1986). For environmentalists, the attractions include lowered contamination of groundwater and surface waters, reduced risk of poisoning wildlife, and enhanced wildlife habitat because of more diversified cropping systems (Cacek, 1984; Papendick et al., 1986).

These benefits for the rural environment also explain part of the appeal of alternative agriculture to agrarian organizations. But their interest goes further. If farmers can use on-farm resources more effectively, they will be less dependent on input suppliers, such as agricultural chemical manufacturers. Agriculture will thus become more agrarian and less industrial in character, reversing a trend that some farmers' groups have decried. Some groups see a reduction in input use as directly benefiting the structure of agriculture. However, the notion that alternative agriculture will "save the family farm," for example, seems questionable (Buttel et al., 1986; Buttel, 1992).

The idea that farmers should cut down on purchased inputs got a strong boost from energy price increases and threats of energy shortages in the 1970s. The problems were especially significant for fertilizers, which accounted for about one-third of U.S. agriculture's energy consumption (Federal Energy Administration, 1976). Energy problems not only increased short-term production costs but also caused serious concern over agriculture's long-term dependence on industrial inputs derived from nonrenewable resources (Breimyer, 1978). Less publicized problems with potash and phosphate in the 1970s reinforced this concern.

THE SOIL CONSERVATION MOVEMENT

Alternative agriculture also owes a debt to the soil conservation movement. The two are closely related in their goals and values, but soil conservation has a longer history and a better-developed institutional base. Soil is a preeminent example of a limited, irreplaceable agricultural resource. Besides damaging a resource essential for future productivity, erosion also contributes to current environmental degradation, especially water pollution.

In the 1930s, when the soil conservation movement took off, it not only aimed at increasing farmers' use of conservation techniques, but also promoted stewardship values. In stressing a long-term perspective, the movement anticipated the rhetoric of alternative agriculture by several decades. Moreover, it attached considerable importance to building up soil organic matter (Flach, 1990), which also is a cardinal principle of organic farming.

The relationship between soil conservation and alternative agriculture is not so simple, however. In the post–World War II period, the scope of conservation narrowed. It focused more on specific techniques to control erosion and less on improving soil quality. Conservation tillage often requires greater use of herbicides. As a result, some people fear that soil conservation is incompatible with reduced-chemical alternative systems. However, the incompatibility probably can be resolved through new conservation techniques; organic farmers, who use no herbicides, generally conserve soil well (U.S. Department of Agriculture, 1980).

This technical issue aside, it is more significant that many soil conservationists have recently embraced broader environmental and resource conservation goals similar to those of alternative agriculture (Cook, 1985). This broadening of interest revives the thinking of early soil conservation leaders with an ecologically informed vision, such as Aldo Leopold and Hugh Bennett (Meine, 1987).

Agricultural Ideals

The concept of alternative agriculture that has arisen from these sources is based on several ideals for agricultural systems:

• An agricultural system should have a range of goals and therefore should be judged by various criteria—not just the traditional considerations of productivity and short-run economic returns but also standards for environmental quality, resource conservation, food quality, and the well-being of rural communities. This idea has been promoted by advocacy groups that see alternative agriculture as supporting their social, environmental, and economic aims.

• A farm should be like an organism in the way its parts are harmoniously integrated with each other, with the environment, and with people (Koepf, 1981). It was this connotation, not the chemical one, that the word "organic" originally had in the term "organic farming." It implies that to understand a farming system, one must study it as an entire system, not part by part.

• A farm should be modeled after natural ecosystems; organic farming "takes as its guide the working of biological processes in natural ecosystems. . . . Agriculture is primarily applied biology and is most likely to be successful when it accepts and follows biological principles" (Hodges, 1981, p. 7). Similarly: "A 'correct' agriculture, from an ecologic point of view, should reflect, on a smaller, necessarily simpler scale, the integrated, mutually dependent, symbiotic relationships of coevolved species in a natural ecosystem" (Callicott, 1988, p. 8).

• A farm should use biological principles to reduce its dependence on inputs from sources off the farm (Wagstaff, 1987). This was an important early goal of the organic farming movement. More recently, it caused two different concepts to be combined in the single term "low-input sustainable agriculture."

• Agriculture must be adapted to the specific conditions of each site. This was a precept of the early soil conservation movement, as espoused, for example, by Leopold: "Each acre should produce what it is good for, and no two are alike" (quoted in Meine, 1987, p. 148). Similarly: "It is the infinite variety in land . . . and in other factors of the natural environment that rule out simple, fixed formulas for conservation of soil and water. The old, single-practice methods cannot possibly succeed in meeting the wide range of conditions prevailing across the country, or even within a state or county, or on a farm" (Bennett, 1946, p. 22).

• Successful application of this principle depends on farmers' first-hand observations and understanding. This view was espoused strongly by Sir Albert Howard ([1943] 1976, p. 182) in a book that laid much of the foundation of organic farming.

Implications for Research

These ideals suggest several approaches to improving agricultural research, especially research on alternative agriculture.

1. *More research should be multidisciplinary.* Researchers can then analyze a farm as a system and evaluate alternative systems according to the wide range of criteria listed above:

> The all-important holistic nature of the farm implies interactions between components. . . . These interactions, in the minds of many, limit the degree to which component parts may be meaningfully separated. . . . This line of thought has led to many attempts to research or demonstrate whole systems, with the addition of sociological elements and goals. (Harwood, 1984, p. 2)

Some advocates of this view carry it further (e.g., Callicott, 1988). They argue that research should be based on concepts of knowledge and human understanding radically different from what they say dominates scientific thinking today: "A truly alternative agriculture must be based on a truly alternative science" (Kloppenburg, 1991, p. 542). However, this is not one of the universally accepted core ideas about alternative agriculture. Therefore, we do not propose strategies that involve alternative theories of knowledge. No doubt the way that science views the world should change, and no doubt it eventually will, as it has throughout its history. But progress in agricultural research is not limited by narrow scientific vision as such; it is limited by the narrow way that scientific vision is embodied in actual research, as we discuss in chapter 5.

2. *Research should be grounded in ecological principles.* These principles apply to the entire farming system, which includes not just crops, livestock, and the environment but also humans:

> An agroecological approach . . . takes on a cultural perspective as it expands to include humans and their impacts on agricultural environments. . . . An integration of ecosystem and social system knowledge about agricultural processes will not only lead to a reduction in synthetic inputs used for maintaining productivity, but will also permit the evaluation of such qualities of agroecosystems as the long-term effects of different input/output strategies, the importance of the human element to production, and the relationship between economic and ecological components of sustainable agroecosystem management. (Gliessman, 1990b, pp. 369–70)

3. *Information and farmers' understanding are critical production inputs.* They are as important as technological inputs and must be part of the development of new systems:

> We currently have a generation of researchers, extension specialists, farmers, and ranchers who too often seek to develop and use a "formula" or "product in a package" to solve most production constraints. . . . In the future, we need to stress the importance of eval-

uating a series of alternative solutions to each problem as well as seeking an understanding of the entire production system and environment in which it operates. (Francis, 1990a, pp. 456–57)

4. *More research should be done on working farms.* This will make the experimental conditions more realistic. It also will avoid the oversimplifications of experiment station research, which might be acceptable for studying specific components of a farming system but not an entire system:

> Small plots and farms are very different things. It is impossible to manage a small plot as a self-contained unit in the same way as a good farm is conducted. . . . The plot does not even represent the field in which it occurs. . . . Findings based on the behaviour of these small fragments of artificially manured land are unlikely to apply to agriculture. (Howard, [1943] 1976, pp. 185–86)

5. *Farmers should have a greater influence over agricultural research.* This demand has been raised by some agrarian-oriented organizations of farmers and rural activists. These groups believe that researchers play down farmers' role because of elitist professional values and a desire to promote a more industrial-style agriculture (Watkins, 1990). They want farmers more involved in research so that researchers will benefit from farmers' knowledge and experience, understand better the values that underlie farmers' preferences among production options, and do research that meets farmers' needs:

> The farmers complain that the research workers are out of touch with farming needs and conditions; that the results of research are buried in learned periodicals and expressed in unintelligible language; . . . that the average farmer cannot obtain a prompt answer to an inquiry. (Howard, [1943] 1976, p. 190)

Part 2 analyzes these ideas, for both agricultural research in general and alternative agriculture in particular. Although the ideas seem very promising, sometimes they are turned into oversimplified prescriptions that do not do justice to their complexities. We will try to disentangle some of these complexities. We hope to provide a better picture of when each proposal is or is not appropriate, so that each can make its maximum contribution to improving our agricultural system. To prepare the way, the next two chapters present needed background material: a sketch of historical changes in agricultural research and a delineation of what the term "agricultural research" covers.

3
. . .

A Historical Overview

It would be the grossest injustice . . . to say that agricultural science has never before been prosecuted with zeal, intelligence, and in the spirit of true philosophy.
—Colman, 1856, pp. 6–7

A look to the future of public agricultural research should begin by looking back over its first century. Granted, the rhetoric of agriculture customarily looks forward, emphasizing progress and innovation. We are always, it seems, at the crossroads, on the threshold, at the dawn of a new era. No doubt it is invigorating to think that the problems one is trying to solve are unprecedented. But agricultural issues have a way of recurring, and it would be a mistake to ignore how research has tried to deal with them in the past. At least in a general way, many ideas now being suggested by alternative agriculture supporters were foreshadowed in earlier debates about the research system.

We can read the historical record as a cautionary tale, teaching us that agricultural research moves in a direction determined mainly by external forces. Change has been largely evolutionary and unplanned, a continuing response to dynamic economic, demographic, and technological conditions. Several developments could not fail to affect research strongly: the emergence of the United States as a scientific leader, the great expansion of agricultural input industries, the declining proportion of farmers in the population, and the marked increase in agricultural trade after World War II.

Planned change is further constrained by institutional and professional aspects of the research system itself. These include the differences in prestige attached to different kinds of research, as well as the political implications of research priorities. Thus the public research system has been described (in an article that vigorously praised its achievements) as

> constantly undergoing marginal changes, but these seldom cause deviation from established patterns. . . . The research establishment has a tremendous tendency to maintain the status quo and to continue its lumbering, disjointed movement along well trodden paths. (McCalla, 1978, p. 482)

On the other hand, the record also gives grounds for optimism. Significant and lasting changes sometimes do occur intentionally. The idea might originate within the research establishment, or it might come from outside sources, such as consumer advocates, commodity organizations, or the environmental movement. Outsiders may apply pressure directly on state agricultural colleges and other research institutions. They also may work indirectly through legislation and appropriations at the state or federal level. Sometimes research institutions have responded willingly to such pressure, but often they have gone along only reluctantly (Busch and Lacy, 1983, chapter 11).

To provide a better picture of possibilities for the future, this chapter reviews historical changes in four aspects of agricultural research:

- Administration of the research system;
- Research topics;
- Control of research goals;
- The intellectual and professional setting of agricultural research.

In the past, efforts to change agricultural research have concentrated on the first two aspects listed. The last two, however, are most relevant to this book.

Administration of the Research System

Most public agricultural research in the United States is done in a well-defined institutional system, consisting mainly of the land-grant universities, their associated extension services and experiment stations, and the U.S. Department of Agriculture (USDA). Discussions of change in agricultural research often concentrate on how this system works: planning and coordination, division of responsibilities, financing, and so forth.

This emphasis makes sense. Redefining administrative responsibilities, forming new coordinating and advisory committees, and consolidating bureaus and divisions are obvious ways to try to make research more socially responsive and efficient.

These changes sometimes are dictated by federal legislation. The major farm bills that revise the agricultural commodity programs every few years usually deal also with research. Other changes occur at the initiative of the system itself, which has regular mechanisms for program review and planning. (A good example is a report by the Experiment Station Committee on Organization and Policy [1985], especially the major section, "Institutional Relationships: Process and Product.")

Unfortunately, reorganization often is transitory, with one change sometimes undoing its immediate predecessor. One commentator bluntly put it this way: "As in a recurring nightmare, every effort thus far to integrate public agricultural research institutions has proven unsuccessful" (Hadwiger, 1982, pp. 26–27). A possible explanation of why federal agricultural research "has been disordered by continuous reorganizations, many of them mindlessly destructive" is that "all were motivated by political and bureaucratic forces, not by the needs of science research" (Bonnen, 1983, p. 961).

We mention the institutional system because so much attention has been given to it that omitting it might seem puzzling. However, the system-wide administration of agricultural research is only tangential to the major themes of this book.

Research Topics

THE TRANSIENT NATURE OF RESEARCH AGENDAS

Another common way for the system to change course is to assign new priorities to different research areas. As with administration, the effect may be short-lived. The priorities sometimes are expressed in vague, general goals or "initiatives" that never get embodied in programs. Even if they are adopted seriously, they may be overturned the next time the issue comes up.

The research establishment's priority-setting exercises have been described—by a not impartial critic, to be sure—as "pedantic and predictable" and "less than exciting" (Hightower, [1973] 1978, p. 66). On the other hand, change does not come only from top-down directives. Individual institutions sometimes adapt to new social pressures in meaningful and lasting ways.

LANDMARKS IN LEGISLATION

Besides the priorities set by the research system itself, research topics are specified also in legislation. This influence, too, tends to be transitory. The effect may last only until a perceived "crisis" has passed or until disenchantment sets in (Madden, 1986).

But legislation occasionally does move research in a new direction. Particularly relevant to alternative agriculture are four laws enacted between 1925 and 1977 that extended the scope of research beyond the production orientation that dominated in the early days. In a way, these laws anticipated the calls heard today for more attention to off-farm socioeconomic effects, system-level agricultural principles, innovative agricultural marketing, and food quality.

The *Purnell Act* (1925) established sociological and economic research at the state experiment stations. This act reflected the realization, born of the severe agricultural depression of the early 1920s, that improving production efficiency would not automatically improve rural well-being. But the act did not overcome many researchers' reluctance to tackle agricultural policy issues, and its supporters strongly criticized their inadequate response (Hardin, 1955, pp. 150–53; Danbom, 1992). A half-century later, the matter still had not been addressed satisfactorily. An influential report (National Research Council, 1972, pp. 384–85), popularly known as the Pound Report, criticized "an excessive orientation toward production agriculture in USDA's definition of problems," the "tendency to define development simply in terms of increasing aggregate income of an area," and the inattention to "social consequences of technological change and the impact of a highly efficient production agriculture on the quality of life . . . for rural residents." Supporters of alternative agriculture also give high priority to these issues.

The *Bankhead-Jones Act* (1935), besides expanding socioeconomic research, provided for research into "laws and principles underlying basic problems of agriculture in its broadest aspects" (sec. 1), specifically mentioning conservation of land and water resources. This act foreshadowed ideas on a "holistic" approach to research and the importance of evaluating the social consequences of new technologies. The language that Secretary of Agriculture Henry Wallace (1936, pp. 82–83) used to explain the thinking behind the act sounds remarkably familiar:

> For theoretical convenience, we separate the work into watertight compartments, and strike off boundary lines that have no counterpart in nature. . . . The field of science is social as well as technical and includes the human application as well as the discovery of scientific facts and principles. The scope of science is life as a whole, and not just certain limited aspects of life. . . . Science is a living thing fashioned of many elements, each standing in a dynamic relationship to the whole and having no meaning apart from its place in the pattern. After the analysis of problems, by separate study, there must be a synthesis of the results, a synthesis which tends to grow wider and more comprehensive as the need develops for conceiving the application in terms of social welfare.

The *Research and Marketing Act* (1946) directed USDA to do research on agricultural marketing and distribution. The severe agricultural depression of the 1920s had occurred because production failed to adjust to the reduced postwar demand. The new research was intended to avoid a repetition after World War II by finding new outlets for agricultural commodities and by improving marketing efficiency.

Although the circumstances in 1946 were different, this work has a counterpart in present-day alternative agriculture research, especially on systems that produce differentiable products, such as certified organic, or that involve more direct producer-consumer linkages, such as farmers' markets. However, the 1946 act had a strong commodity focus. It was heavily oriented toward distributors—so much so that it did not even use the word "farmer" (Banfield, 1949, p. 62). That is why its authors rejected the alternative of allowing commodity prices to fall, with farmers diversifying to reduce production of crops that were in surplus (Banfield, 1949). In contrast, current interest in alternative marketing does not shift the focus away from production. Instead, the goal is to expand farmers' marketing role to let them capture more of the retail value of the product.

The Research and Marketing Act also paralleled present alternative agriculture research in establishing advisory committees with members drawn from commodity organizations, cooperatives, and other nonresearch groups. These committees advised on research priorities and suggested specific research topics (Banfield, 1949; Mainzer, 1958). Similarly, alternative agriculture research programs commonly give more influence to research "outsiders," especially farmers. This trend has revived questions that surrounded the earlier committees (Hardin, 1955, pp. 102–6). Are they purely advisory, or is their "advice" binding? Should they deal with individual projects in detail or only with broad priorities? Will the process give unfair leverage to certain interest groups while excluding others?

The *Food and Agriculture Act* (1977), our last example, directed U S D A to undertake research in human nutrition beyond its existing home economics program. This law reflected the idea that "nutrition and health considerations are important to United States agricultural policy" (sec. 1421a). Specifically, it called for research on "the nutrient composition of foods and the effects of agricultural practices [on it]" (sec. 1422). So, too, alternative agriculture typically gives more attention to the nutritional and food safety aspects of production systems. This contrasts with an earlier attitude that separated such considerations from production-oriented research.

Control of Research Goals

THE POLITICS OF RESEARCH AGENDAS

Choosing the research agenda is sometimes treated as an administrative matter. Thus the research titles of recent farm bills have established various advisory boards and committees (for example, the National Agricultural Research and Extension Users Advisory Board, set up under title 14 of the Food and Agriculture Act of 1977). Typically, these laws specify the constituencies that must be represented, such as consumers, food marketers, and environmentalists. Their rationale is that developing sound research policy mainly means drawing on the various groups who have ideas to contribute.

Such advisory groups may deal with research priorities in a bland, noncontroversial way. However, certain changes can occur only if one takes a frankly political view. The groups who want to move agricultural research in various directions often are separated by deep-seated conflicts and antagonisms, not just by differences in perspective.

For example, nutrition was included in the 1977 Food and Agriculture Act largely because of agitation by "new agenda" groups, including the environmental, consumer, and rural activist organizations mentioned in the preceding chapter (Paarlberg, 1980, chap. 5; Hadwiger, 1982, chap. 8; Stansbury, 1986; Kerr, 1988). The legislation was a significant victory for people who had sought to open the process of deciding research priorities. Previously, this prerogative had fallen largely to the agricultural research establishment and a narrow spectrum of outsiders, such as commodity groups, the major farm organizations, and agribusiness trade associations.

Similarly, current interest in greater involvement of farmers in research is only partly a matter of making research more effective by drawing on their ideas. Its challenge to researchers' exclusive right to choose the agenda is a political issue, too.

CONTROL OVER RESEARCH IN THE EARLY DAYS

That challenge is a new twist in an old debate. Control of the research agenda has been controversial from the start of the research system, but at first the issue was framed differently. Farmers were skeptical about experiment station research, and some opposed it strongly (Danbom, 1986b; Marcus, 1986). Most did not want a greater say in the research agenda—they wanted no research agenda in the first place. Instead, they preferred that the stations stick to their service role, such as exposing fraudulent claims for fertilizers ("fertilizer control").

Early station directors had other ideas. Speaking shortly after the nation's

23

first agricultural experiment station was established, its director offered the following response to a call for broad research on such topics as soil-plant relationships, weeds, tillage, and animal digestion:

> It would be perhaps more for the interests of the State of Connecticut
> . . . to attempt, for instance, to answer some of the questions . . . pro-
> posed here . . . than even to continue the fertilizer control. There is no
> difficulty in undertaking such investigations and in bringing them to
> successful issue if the farmers of Connecticut will say, "we want it
> done, and here are the means to do it with." (Johnson, 1882, p. 56)

To gain support for experiment stations as true research institutions, the most successful station directors (called "research-entrepreneurs" by Rosenberg [1971]) at first steered their stations' research toward practical knowledge that could be used immediately by farmers (more precisely, by the more commercial farmers). Later, their scope would broaden to include less immediately applicable research on scientific principles. The shift from service to research was expressed formally in the Adams Act (1906). It received further support with the establishment of the Cooperative Extension Service (the Smith-Lever Act of 1914), which freed station scientists from having to do extension work.

Once the scientific mission of the research system gained general acceptance early in this century, debate over controlling the research agenda largely subsided:

> By 1900, the agricultural research establishment had developed a habit
> of mind which made critical self-analysis difficult. Concentrating on
> increasing production, researchers failed to see the potential validity
> of other goals . . . [and saw] their critics as backward, reactionary im-
> pediments to social progress. (Danbom, 1986b, p. 117)

Even during the upheavals of the Great Depression, which shook up so many areas of agricultural policy, "the traditional orientation of the [re-search] system was too deeply embedded in its very purpose, too habitual, and too heavily reinforced by support groups to be changed quickly or eas-

24

ily" (Danbom, 1986b, p. 120). Only recently have researchers' obligations to farmers become a significant issue again. A century ago, the stations had to push the idea of research on farmers; today, farm groups are pulling on the stations to make them do the kinds of research farmers want.

THE "NEW AGENDA"

In contrast to production-oriented research, agricultural social science in the 1930s underwent unprecedented intellectual ferment and creativity. At first, the change seemed only temporary, because social issues would soon be pushed aside by the economic relief provided by World War II. However, the issues were revived by the "new agenda" groups of the 1960s and 1970s, when earlier faith in the social value of scientific progress was giving way to ambivalence if not outright disenchantment. The change affected many areas of science, but it was particularly important for agriculture because of concern about environmental damage by pesticides. This concern, aroused so effectively by *Silent Spring* (Carson, 1962), was a leading contributor to the reawakened environmental movement.

Another significant contributor to the new agenda was the greater attention being paid to the lower classes, including the rural poor, farm workers, and tenant farmers. A generalized radicalism, arising especially from the civil rights movement and the Vietnam War, brought under criticism any economic or political institution perceived as perpetuating social injustice and economic inequality: "There came a change in the national mood that challenged almost anything that was long established" (Paarlberg, 1980, p. 59). This change strongly affected agriculture because urban people, who by then were an overwhelming majority, began to pay attention to agriculture as never before. Their unprecedented challenges to long-established agricultural policies ended agriculture's privileged status in governing its own affairs, insulated from broader social and political forces (Paarlberg, 1980, chap. 5).

The most famous expression of this development was *Hard Tomatoes, Hard Times* (Hightower, [1973] 1978). In keeping with the adversary stance of many new agenda groups, Hightower did not expect much from administrative formalities like coordinating committees and user advisory boards. His highly publicized and controversial book, whose spirit continued a long tradition of populist agitators, dealt with control over research as a matter of raw political and economic power. That power, he argued, had been concentrated in the hands of agricultural input suppliers, processors, and corporate and other large-scale farmers, to the detriment of family farmers, tenants, farm workers, consumers, and the environment.

The book was a harbinger of today's alternative agriculture movement. The connection is not so much that it briefly mentioned how research had failed to consider alternatives to agricultural chemicals. More important is that it called for agricultural research to serve a broader clientele. Its agrarian tone is echoed in alternative agriculture rhetoric today, especially among grass-roots farmer and rural activist organizations that demand a greater influence on research (Burkhardt, 1991).

The major federal initiative in alternative agriculture research, USDA's Low-Input Sustainable Agriculture program (Madden and O'Connell, 1990), resulted directly from agitation by such groups. When the idea of "alternative" or "sustainable" agriculture attracted considerable support in the early 1980s, USDA opposed legislation directing it to do research in this area. Such research nevertheless was mandated by the Food Security Act of 1985 because outside groups had become powerful enough to insist on it.

The Intellectual and Professional Setting of Agricultural Research

Research administration, research topics, and control over research goals lend themselves to specific intervention, such as a law that establishes a new program. Less tangible, but still important, is how research gets done. After research topics and administrative arrangements have been chosen, many issues remain, such as research quality, the balance between "basic" and "applied" research, and the role of disciplines.

THE EMERGENCE OF DISCIPLINES

An important early development was the formation of specifically agricultural disciplines, such as agricultural economics, agronomy, and dairy science, by the first quarter of this century (as detailed in chap. 5). These disciplines established the prevailing modes of scientific communication, set standards of quality, and laid down the boundaries of a scientist's work. (Currently, attempts are being made to establish a new discipline, agroecology, as the scientific underpinning of alternative agriculture; see chap. 6.)

The establishment of the agricultural disciplines signaled a shift in the orientation of agricultural research. Under a disciplinary orientation, research questions are chosen more for their potential to advance disciplinary understanding than for the possibility of solving an existing problem. Not all "real world" problems are well matched to prevailing disciplinary boundaries and concepts.

In agriculture, the "discipline first" idea was severely challenged by the Great Depression, especially among social scientists. The dreadful state of the farm economy showed that a laissez-faire policy did not work. Because the New Deal's innovative planning activities in land use and other agricultural and rural issues were so well regarded, social scientists could retain their legitimacy while getting involved in action programs (Taylor, 1947; Kirkendall, 1966, chap. 12).

This represented a temporary return to the social sciences' original character in the mid-nineteenth century (Geiger, 1986, pp. 25–29). An influential advocate of social scientists' involvement in action programs, rural sociologist Carl Taylor, argued that applying sociological theory to social problems would strengthen the discipline by letting it test its theories empirically, since "social laboratories are only to be found where social action is in process" (Taylor, 1941, p. 159). Unfortunately, this vision seems not to have prevailed among agricultural social scientists. The competition—real or perceived—between a disciplinary and a problem-solving orientation is still with us, and instead of discussions of how to combine the best of both, we more often get arguments over which is more legitimate (Madden, 1986).

COMPLACENCY AND UPHEAVAL IN THE POSTWAR ERA

In the prosperity of the post–World War II period, agricultural research put aside its social agenda and again took its main goal to be increased agri-

27

cultural productivity, which it sought to achieve primarily through discipline-based research. It was so successful that there was not much reason for critical self-examination (Danbom, 1986b; 1992). Researchers devoted little attention to what their job should be, how they should go about it, or how well they were doing it.

This complacency was shaken by two reports issued by prestigious panels made up mostly of research establishment insiders. The first, the Pound Report (National Research Council, 1972), received wide publicity because it condemned the "shocking amount of low quality research in agriculture" (p. 70), which it characterized with words like "outmoded," "pedestrian," and "ineptness" (pp. 11–12). Among its wide-ranging recommendations was a call for closer ties between agricultural and related basic science disciplines. Noteworthy for alternative agriculture, one of the reasons for this recommendation was to help agriculture deal with environmental problems (pp. 56–57).

A decade later, another review of the status of agricultural research (Winrock International Conference Center, 1982) also evoked a strong reaction from the research establishment. Like the Pound Report, this review recommended more competitive grants.

A new funding mechanism might seem like an administrative matter of the kind we previously downplayed, but it can have important consequences. Traditionally, the federal government has supported the state experiment stations through block grants allocated by formula. The small program of competitive grants mandated by the Food and Agriculture Act of 1977 was a significant alternative to formula funding. These grants were not restricted to scientists at agricultural experiment stations. They also were available to scientists at nonagricultural institutions who did fundamental research in areas related to agriculture. Conversely, agricultural researchers who had not previously done fundamental research were encouraged to do so, since the grants covered basic processes such as photosynthesis (Huang, 1988).

The Winrock panel favored expanding the 1977 competitive grant program in the hope of ending the isolation of the specialized agricultural disci-

28

plines, a problem that the Pound Report had bemoaned. The argument was repeated a few years later when the National Research Council (1989a), arguing for a substantial increase in agricultural research funding, suggested competitive grants as the main way to allocate it; that recommendation was incorporated into the 1990 National Research Initiative (U.S. Department of Agriculture, 1992).

Drawing on the Past

This review has highlighted changes in agricultural research that have a special bearing on the following chapters. What, if anything, does the historical record tell us about the possibilities for changing research in the ways suggested by alternative agriculture?

As we said at the start, the question can be answered either pessimistically or optimistically. Agricultural research has undergone great changes, to be sure. But much of the change has occurred because of external forces— advances in other areas of technology, developments in international markets, demographic shifts, and so forth. Often the results have been unconscious and unplanned, with agricultural research on the receiving end. We have emphasized intentional changes. Taken together, these changes seem formidable, but by comparison with the things that "just happened," conscious intervention often seems marginal.

We can draw encouragement from a striking exception that is the biggest change of all: the existence of the public agricultural research system. That change decidedly was not a natural consequence of external conditions. It was conscious and purposeful. Most encouraging, it was visionary. Given the primitive state of American science in the late nineteenth century, there was little objective reason to expect the new system to pay off so handsomely. Yet even after we set aside the self-congratulatory rhetoric that often accompanies the system's accounts of its own history, and even after we acknowledge that in solving some problems the system has created others, there is no question that the country and the world have been well served because agricultural research got such strong public support so early.

Does that still matter? That far-sighted action happened a long time ago— too long ago, perhaps, to be very inspirational to people grappling with today's problems. What about the other developments described in this chapter, especially those closer to our own time? Do they offer any help to those who want to apply the ideas of alternative agriculture to research of the future?

We believe they do. Granted, the historical record by itself is not a concrete guide to action: it must always be interpreted in relation to current circumstances. Still, by offering examples of strategies that have worked and others that have not, this large body of experience can inform future efforts to bring about deliberate changes.

4

. . .

What Is Agricultural Research?

All branches of agriculture . . . show a great difference of opinion as to what constitutes research, and mark every gradation from isolated experiments of purely practical import to investigations of the most abstract character. . . . There is often difficulty in determining whether a given topic is an original one or not. . . . Investigations in connection with which there is good reason to expect the establishment of principles of broad application should be preferred to those which have only local or temporary importance or from which only superficial results are to be obtained.
—Committee on Experiment Station Organization and Policy, 1907, pp. 76–77

Before we discuss changes in agricultural research, we want to clarify what we mean by the term—not only what gets studied, but also how, why, and by whom.

We use the term broadly, encompassing some activities that others might exclude. The reason is that agricultural research covers a wide range of phenomena. Moreover, its results are used by different kinds of people—other researchers, farmers, extension workers, input manufacturers, and so forth—and there are many paths leading to practical benefits. This diversity makes it hard to delineate the boundaries of agricultural research in a simple, compact statement. But it also is a source of vitality.

At least it should be. Unfortunately, diversity has not always been viewed in that way. Many kinds of research have something to contribute, but discussions of research strategies do not always allow for all the possibilities.

Instead, we get universal pronouncements about how agricultural research must be done.

Agricultural researchers have varying scientific backgrounds and work in different institutional settings. It is understandable, therefore, that they have varying views of so broad a field. The title question of this chapter is likely to be answered in different ways by professional researchers, farmers, and other groups interested in research. Furthermore, researchers' answers will be influenced by whether they work at an agricultural college, a non–land-grant university, a nonprofit organization, or a government laboratory. (Different, too, will be the perspective of researchers in industry, although this book does not deal with proprietary research.) The answer also will vary among researchers in the so-called "hard sciences," production-oriented fields, and the agricultural social sciences. Forming a comprehensive and balanced view is made even more difficult by the fact that agricultural research is not an area of disinterested inquiry. Its results have important social, economic, and political implications; even its definition has a political side, as we will see when we discuss farmers' role in choosing research topics (chap. 9).

Not surprisingly, therefore, the value of some kinds of agricultural research can get overlooked. The definition of agricultural research should give due recognition to each kind, so that none can impose inappropriate restrictions on any other or deprive any other of its fair share of resources and prestige. That is why in part 2, when we discuss different approaches to agricultural research, we rarely offer rigid dicta. For example, some people would not include an on-farm study under research; our definition admits it provided that it addresses a question appropriate for research. Similarly, part 3 emphasizes flexibility in organizational and policy issues. For example, in judging the value of a researcher's work, we call for more flexibility than is common in many institutions and disciplines.

We do, however, impose one important restriction: that research contribute to cumulative knowledge and understanding. That is why we take a more limited view than some others of how closely research topics should be matched to farmers' immediate needs and specific production problems. On

32

the whole, though, we interpret the term "agricultural research" liberally, to allow diversity to flourish.

The Twofold Character of Agricultural Research

Agricultural research has great diversity because its character comes partly from agriculture, partly from research. It is like many other fields of scientific research in the ways its practitioners view the world and in the methods they use. Also, its objects of study—plants, animals, microorganisms, weather, soils, and so forth—are among those we are accustomed to study scientifically.

However, agricultural research has its own special flavor. First, its research questions come from an important human activity outside the world of science. Unlike the traditional "pure" sciences, it is not free to set its boundaries as it wishes, as in the glib but serviceable definition of physics as "what physicists do."

Second, agricultural research is supposed to *serve* agriculture, not just analyze it. Its results get put to work in a tangible way by an important industry. This distinguishes it from fields that deal with matters beyond direct human intervention.

Finally, agriculture occurs in commercial production enterprises. It is affected by powerful social, economic, and political institutions, from the local farming community to the global economy. Moreover, farming is more than a matter of technology and economics; farmers have strong traditions and values. Clearly, even though agriculture is based on natural processes, agricultural research cannot take a purely natural science approach.

For all these reasons, agricultural research is less autonomous and self-contained than many other areas of scientific research. This complicates how it should be organized and administered. Both its generically scientific and its specifically agricultural characteristics are essential, but depending on one's background and perspective, it is easy to push one or the other aside. An observation about universities in general applies to agricultural research institutions in particular:

33

The question is not whether diversity should be renounced in favor of any one academic ideal . . . but rather what degree of institutional diversity is essential in order to meet the enormous diversity of needs of our differentiated society. . . . Predispositions for work bench over library, for theorizing over gadgeteering, for naturalist observation over dynamic analysis, for dealing with people over dealing with concepts . . . must all be honored instead of being frustrated in a common mold purporting to serve them all—and matching none. (Weiss, 1964, p. 1199)

Because agriculture encompasses such varying phenomena, many areas of science have agricultural implications—sometimes only indirectly. Although we take an inclusive view of the scope of agricultural research, it still is important to delineate its boundaries, if only imprecisely. To decide whether a piece of research is sufficiently specific to agriculture to count as "agricultural" research, we suggest the following test: Might the work have been done anyway, even if the organism or process it studied were not agriculturally important? For agricultural research as we define it, the answer would have to be "no."

Many fundamental processes, such as photosynthesis, nitrogen fixation, allelopathy, and the hydrological cycle, are interesting and important for many reasons, both theoretical and practical, agricultural and nonagricultural. Thus they are studied by a wide range of scientists, including biochemists, plant physiologists, ecologists, and meteorologists. Much research in these areas would have been done even if agriculture were not important in the United States, and even if we did not believe in research as a way to improve our agriculture. Although some of that work might eventually prove significant for agriculture, it is not what we have in mind when we speak of "agricultural research."

This limit corresponds roughly to the institutional organization of public agricultural research. That is, most of what we would call agricultural research is done in specifically agricultural colleges or research facilities, supported by USDA or other specifically agricultural agencies. Most people doing the work would call themselves agricultural researchers besides identifying themselves by their particular field, such as plant physiology. In contrast, research that has enough other justification regardless of its agricultural significance typically is done elsewhere, by different people.

Applied versus Basic Research

Even when limited in this way, agricultural research has such a broad scope that it is useful to distinguish different kinds. A common distinction is between "applied" research on one hand, and "pure," "fundamental," or "basic" research on the other.

However, under our definition, all agricultural research is applied: the reason for doing it is that it has implications for agriculture, which is an important human activity. Granted, some agricultural research, especially if it deals with general processes rather than specific production methods, investigates phenomena similar to those studied in the indisputably pure sciences, using similar techniques. Granted, also, that the pure sciences have contributed greatly to agriculture. For example, in his fundamental research on plant inheritance at the beginning of this century, George Shull used corn because it was a suitable organism for his purposes (Mangelsdorf, 1951). Unexpectedly, that work led directly to the development of hybrid corn, an outcome described as "an outstanding example . . . of the influence of theoretical scientific research in revolutionizing the production practices of an agricultural crop" (Jenkins, 1936, p. 468).

Despite such connections, we think it is important to distinguish research whose main rationale is its expected value for agriculture from research that might contribute to agriculture eventually, but only indirectly and unforeseeably. Letting an agricultural research area call itself "pure" would give it too much independence from the "real world." It would let researchers define research needs according to their own preferences, or those of their discipline, not according to society's needs.

The label "applied" therefore imposes an important constraint. What is our justification for imposing it on *all* agricultural researchers?

We have already noted that most public agricultural research is done in specifically agricultural institutions, with funding intended specifically for

35

agriculture. Presumably, when agricultural researchers accept the advantages of working in a field with this preferred status, they also accept the goals that led society to bestow that status. The professional reward system sometimes encourages researchers in applied disciplines to abandon their orientation toward external problems and instead to emulate the pure disciplines (see chap. 11). We believe, however, that even when researchers use the techniques of the pure sciences, they remain obliged to work on actual problems in our agricultural system (Bird, 1991, p. 29).

Also, calling some agricultural researchers "pure" scientists might absolve them from having to think about the possible economic, social, or environmental consequences of their work. They should not be allowed to say, "We just do the research—it's up to other people how it gets used." Even in the pure sciences, such an attitude at best is arguable: witness the profound soul-searching that some physicists went through after the atom bomb was dropped. The attitude of limited responsibility is hardly defensible in agricultural research, where the practical implications are so direct and obvious and where we have a record of new technologies that caused serious environmental or social harm.

The Relationship of Research to Farms and Production Systems

For this book, rather than distinguishing between "basic" and "applied" research it is more useful to characterize research according to how it relates to farms and production methods. Some research deals with farms and the actions of people on farms; the questions it asks have no meaning except in the context of agricultural production systems. Other research also investigates processes and organisms found on farms, sometimes exclusively on farms. However, it need not take into account that agriculture is a production process and that a farm is an economic and social entity. ("Applied" versus "not so directly applied" would be an accurate, if verbose, way to distinguish the two types.)

This distinction might seem so obvious that it hardly needs to be elaborated. But too often it gets overlooked, so that a single notion of what agricultural research is and how it should be done gets applied in every case. In

36

parts 2 and 3, we try to undo such oversimplifications. We stress the need to accommodate diversity in the choice of research site, for example, and in farmers' role in choosing research topics. Often, a useful guide to the best choice will be whether the research is specifically about farms and production systems or about agricultural processes.

EXAMPLES OF DIFFERING RELATIONSHIPS

We illustrate the distinction with pairs of related studies from three fields:

Crop science. The first study examines how the yields of different corn varieties respond to drought; this makes sense only within a corn production system because a yield measurement implies a choice of planting density, fertilization, pest control, and so forth. The second investigates the genetic basis of corn's drought resistance, a question that is independent of the production system. A study like this could be done on any species, cultivated or wild, but this study is done on corn because corn is important in agriculture. Moreover, even though the study is not connected to any production system now, its findings may change corn production in the future.

Soil science. One study examines how different tillage methods affect soil physical properties, such as pore size. The second analyzes how pore size affects percolation of water through soil. The first is done with tillage methods that farmers now use or might use in the future; it measures the consequences of a production technique. The second study does not involve a production system. A similar study could be done anywhere, but in practice it would be done on agriculturally important soils because it deals with a phenomenon that is significant for agriculture.

Entomology. The first study measures how cotton yield varies with the abundance of some insect pest; again, because it involves yield, this question is intrinsically tied to production systems. In contrast, the second study examines how the insect metabolizes an insecticide used on cotton. This work can be done without any reference to cotton production. However, the reason for

37

doing it is that the pest and the insecticide are important in cotton production, and its conclusion could improve the production system. For example, it may suggest ways to slow the pest's development of resistance to the insecticide.

"CLINICAL" AGRICULTURAL RESEARCH

To make clearer the significance of this distinction, we draw an analogy with medical research, where the distinction is formally recognized. The first type of agricultural research is similar to clinical studies. A clinical study develops or tests a medical treatment. Often, it investigates not only the effectiveness of some treatment, but also its side effects, contraindications, and cost; these all have counterparts in research on agricultural production methods. A clinical study also may take account of some distinctively human factors, such as how well patients follow a treatment regime; so too, researchers should consider how well farmers will use a new production technique. A clinical study might be done first on laboratory animals, perhaps because of uncertainties about safety; however, an animal study would be intended as preliminary to a study on humans. Likewise, experiment station studies of production techniques complement those done on farmers' fields.

In contrast, some biomedical studies do not involve any medical treatment. Rather, they investigate processes that must be understood if we are to develop better treatments, such as why a cell may start dividing out of control, or how the immune system identifies something as "foreign." The experimental "animal" may be a human to make the results as applicable as possible to better medical treatments for humans. However, the human characteristics of the experimental subject—values, knowledge, perceptions, and so forth—are not relevant: the research deals with human genes, cells, tissues, or organs, but not with people. A similar study might be done on another species, referred to as an "animal model" of a human disease. In all such respects, the research is similar to the second member of each pair of agricultural research studies previously described.

OTHER WAYS TO CATEGORIZE AGRICULTURAL RESEARCH

Superficially, it might seem that we simply are dividing agricultural research into "applied" and "basic" types. However, the distinction we are making

38

had already been made when formal agricultural research of any kind had barely begun in the United States, with nothing that even remotely could be called "basic." An early researcher advocated "a separation of agricultural scientific study into two sections," one consisting of "field-experiments with various crops and with rotations, as well as experiments on the feeding of stock," the other involving research on "the transpiration qualities of plants; on rainfall, evaporation, and percolation; on nitrification; on the sources of the nitrogen of vegetation," among others (Sturtevant, 1882, pp. 40–41). He commented that questions of the former kind are likely to be asked by farmers, whereas the latter are likely to be asked by researchers. He noted that although both deal with similar subjects, only experiments of the first kind "furnish information for the actual carrying out of profitable farming" (p. 49).

Nor are we drawing a distinction between "science" and something else. That a study is about actual production systems, actual farms, or actual farmers does not say whether it is more or less scientific than one about an underlying process. Assuming they are done appropriately, all the studies in our examples deserve to be called "science." Some people restrict that word to research that studies certain phenomena or that uses certain techniques. In contrast, we use the word "science" to mean a way of thinking and of organizing knowledge that applies to many different topics and many different research methods.

When Is It Research?

WHY IS IT EVEN NECESSARY TO ASK?

So far, we have concentrated on the agricultural part of the term "agricultural research." We now turn to the other half of the definition: What makes an activity research?

Researchers in most fields need not think much about this question. In agriculture, however, it still causes considerable disagreement, as it did in the early days of the public research system (Busch and Lacy, 1983, p. 10). In part, the reason is the broad scope of agricultural research. Also, research sometimes is not distinguished from related activities, such as field demonstrations and product testing.

We will suggest boundaries for what research is, but we cannot offer a procedure to determine unequivocally whether a particular activity falls within those boundaries. Inevitably, some cases will require judgment. Nor do we answer the question "Who should decide?" But these operational difficulties do not mean that "research" is an arbitrary or meaningless concept, any more than the difficulties in deciding "Is that object 'art'?" mean that the concept "art" is meaningless. Whatever the problems of definition, it would be hard to discuss agricultural research meaningfully under the principle that "if someone says it's research, it's research."

Unfortunately, proposing limits encounters the problem of prestige:

> Whenever a word becomes used for something considered desirable, useful, or beneficial, it acquires valuable connotations that give it a sort of halo. Many people then attach this word to the names of their objects or activities in order to acquire some of the esteem carried by the word. . . . Research . . . has acquired a halo, because the results of research have been used to make our modern civilization. . . . The unfortunate result, however, is that every study, every investigation, and every search of any kind is called research. (Simons, 1960, p. 80)

When we use the term "research," we do not intend it to have a halo. In saying which activities we include under the term, we are *not* judging their value. To label something research does not imply that it is good; the term for that is "good research."

Conversely, it is lamentable that placing certain activities outside the domain of research is seen as denigrating them. In delimiting that domain, we are not saying whether a given piece of work is or is not worth doing.

LIMITING THE DEFINITION

Our use of the term in some ways is broader but in others more restrictive than some others prefer. We do not impose several criteria that sometimes figure in discussions of research: who does the work, the particular methods used, or the nature of the knowledge sought, that is, whether it is descriptive, qualitative, statistical, theoretical, empirical, and so forth.

On the other hand, we do impose two restrictions that not everyone else accepts. First, we use the term "research" to mean "scientific research." The implication is that the work is done to increase *cumulative* knowledge and understanding; an individual result usually has little value by itself (see chap. 12). Second, the primary "users" of the results must include other people engaged in similar activities, not just the person who does the research.

This narrows the term compared with everyday usage. For example, a stock brokerage might have a "research" department to help it make buy and sell recommendations; a manufacturing company might do market "research" before developing a new product line; someone buying an appliance might say, "I did a little research before picking a brand."

Agriculture, too, involves activities that could be called "research" in these senses. However, we exclude investigations if their results are intended to be used by themselves to improve individual farm operations. Such investigations would better be called "product testing." They include performance measurements on tractors, crop varieties, soil amendments, and so forth. Their goal is to provide information on these products in particular so that farmers can decide which to buy. Such measurements can be worthwhile. But if they report results only for specific items, under specific conditions, without adding to our cumulative knowledge of such products or helping to develop new ones, they should be distinguished from "research" as we use the term.

The same holds for tests of competing production methods, not just products, if the tests are intended merely to solve a particular farmer's production problems. For example, a farmer might run a trial to learn which cultivation technique best controls some weed. (Alternatively, a researcher or extension worker might run it; we have already noted that who does the work does not matter.) Once that weed can be controlled to the farmer's satisfaction, the trial has served its purpose. But unless it can contribute toward solving the problem more generally, it would not come under the label "research" in our sense.

IMPLICATIONS OF HAVING TO CONTRIBUTE TO CUMULATIVE KNOWLEDGE

If research is to fulfill this key requirement, its results must be available to other researchers well into the future: cumulative knowledge builds up over time, not just through simultaneous studies of the same question. Generally, this requirement will not be fulfilled merely by the "fugitive" literature, such as newsletters or reports published informally by the group doing the work. However, these may be valuable as supplements to traditional research publication channels, for reaching readers other than researchers, or for reporting work in progress.

The goal of contributing to broader knowledge requires more than an appropriate medium of communication. Sometimes the research must be planned with that goal in mind. For example, it might seek to delineate the "recommendation domain" for a new practice, to use the language of farming systems research (Shaner et al., 1982, p. 44). The purpose would be to determine where else the results can be applied. Typically, this goal means paying more attention to errors and uncertainties, sources of variation, sensitivity to experimental conditions, and so forth. Contributing to cumulative knowledge also could mean choosing conditions specifically to discriminate between conflicting results from earlier studies of the same question.

Another way to give broader meaning to a particular investigation is to use it to test a theory. The point can be overstated, however. Some people, especially in the agricultural social sciences, automatically reject research that doesn't have a "theoretical framework" or "conceptual model." (We return to this point shortly, when we discuss the varied kinds of knowledge that can be sought under the label "research.")

THE USEFULNESS OF RESEARCH

The desire for results of broad interest can create a dilemma in deciding what kind of investigation is most valuable. There is a tradeoff between studies that we would call research because they strive to contribute to cumulative knowledge and studies that are directly usable by one farmer in particular. The problem is that a good production system must be tailored to the particular farm.

This raises the question "Who are the 'users' of research?" Because contributing to cumulative knowledge is critical, we said earlier that the primary "users" are researchers. This requirement does not mean, however, that research need not also be useful to other people, such as farmers. We have already insisted that all agricultural research is "applied," meaning that it must be useful outside the world of research. But the path from the research project to the farmer is not direct. Typically, the results of one project are used to understand some phenomenon better; the way the results get used by farmers is through the improved practices that emerge from this understanding. Even a field study of a particular production method does not usually give farmers a prescription for solving their problems.

We say this not as a criticism of researchers; it is not intended to support the charge that researchers are neglecting their responsibility to be useful to farmers. For example, in his discussion of different kinds of research that we mentioned earlier, Sturtevant held up the practical field experiments at Rothamsted, England, as admirable examples of research oriented toward farmers' questions. However, he added that "even this superb series of field experimentation seems unfitted to offer the farmer correct reply to what he would know: What manure shall I use on *my* land? How much should *I* use?" (Sturtevant, 1882, pp. 41–42; emphasis in original).

DEFINING RESEARCH BY ITS PURPOSE, NOT ITS METHODS

Having severely limited the definition of research, we now broaden it beyond what some others would accept. Whether work counts as research is not primarily a matter of using methods that are considered standard in the field. We require only that the methods be reasonably likely to achieve the study's purpose. Accepted research methods are usually a good way to meet this requirement, but if they were the only way, there would be no innovative research methods. More important than whether a study uses what is considered *a* scientific method is whether it uses *the* scientific method. That is, does it have reasonable protections against bias and error, is it open to the scrutiny of others, and so forth? A nonstandard method does not prevent a study from being scientific; neither does a standard one guarantee it.

Defining research by the methods used can have unfortunate results. It may make the researchers adopt familiar methods where they are inappropriate, lest the study not be accepted as "valid" research or even as re-

43

search at all. Even worse, the researchers may simply pass over an opportunity to do interesting work because it requires an unconventional approach.

Unfortunately, practitioners of particular kinds of research are sometimes methodological chauvinists. Devotees of small-plot work, for example, may feel that if an investigation involves whole farms, or case studies, or surveys, or interviews, it is something other than "real" research. A related chauvinism is insistence on certain kinds of scientific knowledge and understanding. Typically, this attitude accepts only research that involves a hypothesis, not "mere" description. Thus for federal competitive grants in agricultural research, the first criterion for all proposals is "conceptual adequacy of hypothesis" (U.S. Department of Agriculture, 1992, p. 38). Yet these grants are supposed to cover areas where the notion of a hypothesis is not meaningful, such as "empirical estimates of the impacts of sustainable practices on the competitiveness of U.S. produced agricultural commodities" and "exploratory research to focus on new ways to improve the social and economic well-being of rural families" (p. 20).

The prestige of inferential (hypothesis-testing) statistics may make researchers force their data into the preferred mold when it is not appropriate. This practice leads to false conclusions that get accepted anyway, simply because they have the right form. In a review of ecological field studies in which he looked for just one kind of error—the lack of appropriate replication—Hurlbert (1984) found that 48 percent of studies that used inferential statistics were not justified in doing so. To cure this problem, he advised journal editors to

disallow the use of inferential statistics where they are being misapplied . . . [but] be liberal in accepting good papers that refrain from using inferential statistics when these cannot validly be applied. Many papers, both descriptive and experimental, fall in this category. Because [of] an obsessive preoccupation with quantification . . . it is often easier to get a paper published if one uses erroneous statistical analysis than if one uses no statistical analysis at all. (p. 208)

44

THE ROLE OF DESCRIPTIVE AND EXPLORATORY RESEARCH

We consider it ill advised to limit the term "research" to studies that test hypotheses or that try to establish causal connections. Exploratory, descriptive, and qualitative studies also can be appropriate. Knowledge is not always sufficiently developed for researchers to propose or test meaningful hypotheses. We require only that the mode of research be suitable for significantly advancing our knowledge of some question, taking account of where that knowledge stands at the time. Thus cumulative knowledge could be advanced by research that systematically collects and compiles many pieces of information of a kind that, taken in isolation, might be called "local," "site-specific," "indigenous," or "experiential."

However, collecting exploratory or descriptive information is not always justified. Eventually, enough descriptive information is accumulated that the challenge is to make sense out of it. At that point, collecting more is no longer a sensible way to advance our knowledge.

Unfortunately, each experiment station's mission to serve its own state may prompt it to do a study much like those already done in neighboring states under similar conditions. In alternative agriculture research, which is more advanced in some states than others, exploratory projects may seek nothing more than to "see what's out there," even though colleagues in neighboring states could have given a good answer.

We leave open the question of how to decide whether a study has significantly advanced our cumulative knowledge and understanding. We can say unhesitatingly, however, that if the investigators did not even mention this purpose when giving the motivation for their study or did not address it seriously when reporting their results, the work would not count as research by our definition. Conversely, it seems prudent not to be too restrictive, but instead to accept any bona fide attempt to fulfill what we consider the purposes of research; the term should encompass legitimate failures.

• • •

In the chapters that follow, we examine several research approaches in more detail. We try to keep the discussion applicable to all kinds of public agri-

45

cultural research. However, despite what we have said about diverse styles of agricultural research, the balance may seem tipped toward certain kinds. We talk mainly about research on specific production methods rather than on more general processes, and about research done in the field rather than in laboratories and growth chambers. This focus is in keeping with the current tone of alternative agriculture, which is the source of the innovative research approaches at the heart of this book. Because of its connection with political issues and activist groups, alternative agriculture research often deals with specific systems. Moreover, it is still young: its ideas have not yet had a chance to mature, and its concept of agricultural research understandably emphasizes the tangible and the realistic.

That is the right place to start. As alternative agriculture evolves, presumably it will give full weight to all modes of research, including those that are less directly linked to production systems. Meanwhile, in illustrating our arguments, we emphasize where research stands now, while trying also to keep in mind what it might become.

Part Two

THE PROCESS OF AGRICULTURAL RESEARCH

5

. . .

Multidisciplinary Research

Agriculture . . . in its improved condition, combines so many arts and such various subjects of inquiry and observation, that a close scrutiny and long-continued inquiry are . . . indispensable to a thorough knowledge of it.—Colman, 1856, p. 4

The world of agriculture encompasses an extraordinary variety of organisms and processes: people, their social institutions, and their technologies; plants and animals, both domesticated and wild; microorganisms; and weather, soil, and natural resources. As applied scientists, agricultural researchers must choose their research questions according to the problems occurring in this multifaceted "outside" world; they cannot simply follow wherever their intellectual curiosity leads them.

Of necessity, however, research must organize itself into subdivisions, or disciplines. Only by chance will an agricultural problem fit into a single discipline. True, disciplinary boundaries change over the years, but not nearly as rapidly as does agriculture. Thus we cannot hope to divide research into domains that correspond precisely to agricultural problems.

This leaves researchers a choice: they can change the research questions to fit their disciplines, or they can override disciplinary boundaries to deal with problems as they occur in the outside world. The first approach is easier, but it is not what applied sciences are supposed to do. If we take seriously the responsibility that comes with accepting society's support of agricultural

49

research, agricultural problems must dictate the structure of research, not the other way around.

For this reason, multidisciplinary research is especially suitable for agriculture. Fortunately, increasingly effective challenges have been raised against the dominance of disciplinary research. But multidisciplinary research is a subtle business, not an all-purpose substitute for single-discipline research. In this chapter, we examine some of its implications, in the hope that agricultural researchers can capture the full potential of the multidisciplinary approach by applying it discerningly.

We first present a simple classification of different modes of multidisciplinary research as preparation for addressing several questions:

• What disciplinary structures are best suited to projects with various goals?

• In what ways is multidisciplinary research especially suited for alternative agriculture, where it is promoted especially strongly?

• For agricultural research as a whole, how completely do specialization and single-discipline research dominate?

• Is multidisciplinary research needed to overcome the limits of prevailing scientific thinking?

A semantic point must be clarified first. For simplicity, we use "multidisciplinary" broadly, meaning any alternative to single-discipline research (including *non*disciplinary research). It would be convenient if there were standard terms for different disciplinary relationships; for example, "multidisciplinary" when several disciplines are involved in a loose way, but "interdisciplinary" when there is more interaction among them (Rhoades et al., 1986). But current usage is inconsistent, so we will give these variations ad hoc labels for purposes of this discussion.

Modes of Multidisciplinarity

Disciplines can collaborate in various ways. The following simplified breakdown, which has two dimensions, is offered with the understanding that in reality the types shade into one another (Lockeretz, 1991b).

The first dimension, which is common in classifications of this kind, re-

fers to how strongly the disciplines interact:
- *Additive*: loosely coordinated disciplinary components;
- *Integrated*: more interaction among disciplinary components;
- *Nondisciplinary*: no disciplinary components;
- *Synthetic*: fusion of disciplinary components into a new perspective.

The second dimension refers to whether the various disciplines cover the same or adjacent territory:
- *Extensive:* Disciplines combine "horizontally," studying several portions of a topic.
- *Intensive:* Disciplines focus on the same question to try to understand it better.

LEVELS OF INTERACTION AMONG DISCIPLINES

Additive. In the weakest interaction among disciplines and the easiest kind to achieve, a study consists of disciplinary components that are loosely coordinated. For example, the components take place on the same site at the same time, with the researchers meeting periodically to discuss their results and make plans. Beyond this coordination, each disciplinary component goes ahead almost as if it were self-contained. This relationship is what people sometimes mean by "multidisciplinary," in contrast to stronger terms like "interdisciplinary."

Integrated. The next strongest relationship gives special attention to how the disciplinary components are linked and to phenomena that either overlap or fall between them. The benefits of this integration come at a cost: team members must work to explain their own contribution and to understand and appreciate their colleagues' contributions (or to learn new material themselves). Also, each discipline's intellectual limitations and its cherished but not necessarily valid assumptions are exposed to the scrutiny of people who have not been trained to accept them unquestioningly. Reconciliation of disciplinary viewpoints, which Rhoades et al. (1986, p. 22) "euphemistically call 'constructive conflict,'" can be stimulating, but it also can be painful. However, if the researchers shy away from it, the result will be "lowest common denominator" research (Blackwell, 1955, p. 370).

51

In this kind of study, and even more in the closely knit "synthetic" mode described later, the investigators must be willing to put aside any notions of their own discipline's superiority. If they do not have much multidisciplinary experience, especially if the members of the group have not worked together before, considerable effort may be needed to blend their perspectives and interests into a smoothly functioning team. Some groups have found it advisable to engage professionals who specialize in team building.

Nondisciplinary. In the two previous types, the project is built up from disciplinary components. An alternative is choosing not to have disciplinary components. In this mode, the foremost concern is *what* is being investigated, not *how.*

This approach may be suitable for exploratory work on topics where there are no theories to guide the research. It is the way much of the research on agriculture's energy problems was done in the 1970s—for example, the systematic calculations of energy consumption by state, type of fuel, commodity, and function (Federal Energy Administration, 1976). To be sure, this effort drew on knowledge from several disciplines, such as agricultural economics, agricultural engineering, agronomy, and chemical engineering. However, the topic is so broad that its disciplinary components are not discernible.

In such work, the disciplinary backgrounds of the investigators (or the sole investigator) are not very important: the researchers draw from whatever disciplinary knowledge they need. An obvious inefficiency is that they first must learn about areas outside their own training. On the other hand, because they are not hemmed in by disciplinary limitations in outlook, they may tackle questions that otherwise might not be asked at all.

Synthetic. Another force that causes disciplinary boundaries to disappear is the emergence of new concepts or theories that were not foreseeable in the contributing disciplines (McIntosh, 1985, p. 25). In this mode, several disciplines fuse their explanations of the same phenomenon into a collective answer that goes beyond a simple combination of their separate answers. This synthesis cannot happen when the disciplines are indiscriminately thrown together: "The unity of learning achieved in this way is merely a unity of accumulation, like a heap of stones" (Gusdorf, 1977, p. 588). Only occasionally are disciplines likely to have a strong underlying connection that has not yet surfaced.

An example of the power of disciplinary synthesis is the discovery early in this century of vitamin A, the first known fat-soluble vitamin. This discovery was made possible when developments in organic chemistry became available to scientists doing animal feeding studies. Agricultural experiment stations were investigating the comparative nutrient value of various livestock feeds. However, people in the field believed that the major nutrients—proteins, fats, carbohydrates, and minerals—were all that mattered (Rosenberg, 1976, p. 186). They also thought that all fats were nutritionally equivalent, providing only energy.

Simultaneously, great progress was being made in protein chemistry, which at the time was a branch of organic chemistry, unconnected with human or animal physiology. This progress allowed nutritionists to conduct trials of feeds formulated from purified proteins and other nutrients. When animals failed to thrive on such feeds, investigators concluded that feeds used by farmers must contain a nutrient that was essential despite occurring at a much lower concentration than any nutrient known at the time. Using fats from different sources refuted the belief in fats' nutritional equivalence: the unknown substance was contained in certain fats, such as butterfat or cod-liver oil, but not in others, such as lard or olive oil. (See McCollum et al. [1939, chaps. 2 and 12] for a discussion of these developments by a co-discoverer of vitamin A.)

The discovery of vitamin A represented more than the solution of a particular scientific problem: it was an early triumph in what soon became the new discipline of biochemistry. Research proposals often talk of combining people from different disciplines to synthesize a new perspective, but true synthesis of disciplines is rare.

Agroecology, when defined to include cultural, economic, and political institutions, is a current attempt at such a synthesis (Altieri, 1989; Elliott and

53

Cole, 1989; Gliessman, 1990b). It is not merely a branch of traditional ecology, because agroecosystems are subject to strong human intervention. As a conscious attempt at a disciplinary synthesis, it is still young. In the next chapter, we discuss its attempt to analyze agricultural systems in a comprehensive way that goes beyond what ecology and the established agricultural sciences already offer.

EXTENSIVE VERSUS INTENSIVE MULTIDISCIPLINARITY

Although disciplinary synthesis is an intriguing prospect, a more common goal when combining several disciplines is to make research more extensive by covering more pieces of some topic. For example, in a study of soil conservation, an economist might calculate the costs of various tillage techniques, a rural sociologist might study farmers' attitudes toward them, a soil scientist might measure their effects on soil structure, and so forth. Such "extensive" multidisciplinary studies might be additive, integrated, or nondisciplinary.

The extensive mode does not exploit the full potential of multidisciplinary research. A form that is less common, but potentially very fruitful, is cooperation by several disciplines in trying to *explain* the same phenomenon, rather than describing or measuring more pieces of it. Such "intensive" research might involve synthetic interaction among the disciplines, as in the vitamin A example. It also can be done with a looser connection, that is, in an additive or integrated mode.

Intensive multidisciplinarity is especially suitable when individual disciplines have failed to answer a question. For example, investigators from different disciplines have separately tried to learn why farmers adopt or fail to adopt soil conservation techniques. One approach is grounded in sociological research on the diffusion of innovations (Nowak, 1984). Another view is that farmers' conservation decisions are determined mainly by economics (Seitz and Swanson, 1980).

Since neither discipline can offer a full explanation despite several decades of research (Lockeretz, 1990), both explanations can at least be combined additively. For example, a regression model might have two sets of in-

54

dependent variables, one set from each discipline (Nowak, 1987). A more integrated approach would recognize that farms are simultaneously economic enterprises, places to live, and the building blocks of farm communities. Therefore, it is artificial to separate sociological and economic variables. A "socioeconomic" analysis might be more effective than one that is merely "sociological *plus* economic." But one could go further, focusing on what the two disciplines cannot explain even after their contributions have been combined. The hope would be to reveal previously overlooked factors in farmers' conservation behavior (Lockeretz, 1990).

Matching the Structure of Multidisciplinary Research to Its Purpose
MULTIDISCIPLINARITY AS A MEANS, NOT AN END
The first step in multidisciplinary research is to decide whether this approach is needed. Yet this seemingly obvious question is sometimes passed over:

> Everyone invokes interdisciplinarity; no one dares say a word against it. Its success is all the more remarkable in that even those who advocate this new image of knowledge would often find it hard to define. The appeal to interdisciplinarity is seen as a kind of epistemological panacea, designed to cure all the ills the scientific consciousness of our age is heir to. (Gusdorf, 1977, p. 580)

Blackwell (1955, p. 370) criticized the faddism of multidisciplinary social science research, urging researchers to avoid a "shot-gun wedding of disciplines." In a similar vein is the observation that

> attempts to . . . wave the banner of "interdisciplinary research" as something good in its own right . . . come in periodic waves. . . . [However] when scientists begin with only a "let's work together" ideal, they have great difficulty in finding a research topic that requires the whole-hearted cooperation of every discipline involved. . . . Interdisciplinary research [is] useful and mutually stimulating primarily when a *problem* has been first identified and *then* scientists with knowledge and skills relevant to its investigation are brought together to work on it. (Storer, 1972, pp. 259–60; emphasis in original)

55

Assuming multidisciplinarity is needed, the appropriate form depends on what the research seeks to achieve. Is it developing a new production technique, evaluating an innovation, or exploring general principles that might lead to new techniques? The goal determines the appropriate research style: qualitative or quantitative; descriptive or analytical; exploratory or hypothesis-testing. The goal also determines the right kind of multidisciplinarity and the right personnel; it is not a matter of whether the production system being studied, if any, is "alternative" or "conventional."

Although choosing the appropriate form of multidisciplinary research is not easy, the matter sometimes is treated casually. For example, it may be disposed of merely by stating the makeup of the research team: "Our project is very multidisciplinary: we have an economist, a soil scientist, an entomologist, and an anthropologist" (alternatively: ". . . and even an anthropologist"). Just including the "right" disciplines on the project does not guarantee that it can do the right kind of multidisciplinary research. Conversely, to be multidisciplinary in content and spirit, not every study needs team members from all the relevant disciplines. Sometimes the other researchers can acquire the additional expertise outside their fields.

Interestingly, Sir Albert Howard ([1943] 1976), whose ideas were so influential in the organic farming movement and who advocated a wide-ranging approach when developing new production systems, did not favor multidisciplinary teams. He observed that "the net woven by the team is often full of holes" (p. 195). Instead, he recommended that the study be done by "one investigator with a real knowledge of farming combined with a wide training in science" (pp. 194–95). (In the terminology of this chapter, he was advocating the nondisciplinary mode.) The high status of multidisciplinarity in alternative agriculture can make one forget that it is only a means to an end and that multidisciplinarity in form does not always mean multidisciplinarity in spirit.

EXAMPLES OF RESEARCH GOALS MATCHED WITH
KINDS OF MULTIDISCIPLINARITY

Developing an improved production technique. Such research might not have to be multidisciplinary at all. Even if the technique is being developed to be part of a larger system, innovation may begin at the component level; ultimately, however, what counts is how the technique performs as part of the new system. Past efforts to develop new production techniques often have begun with a single discipline; no doubt this approach will continue to serve well.

Screening of farmers' innovations. If the purpose of a study is to learn about promising innovations developed by farmers, individual disciplines are far less relevant. Instead, the best approach may be nondisciplinary. The point is not to overlook promising alternatives even though they may be unconventional and even though the farmers who use them may explain them differently than researchers.

For example, research on organic farming in the United States was held back by agricultural experts' belief that it could not be economically competitive and by its unscientific and countercultural image. Now, however, it is receiving serious attention, largely because some farmers have been using it, apparently with reasonable success.

Evaluations of new methods. When an innovative method, whether developed by farmers or researchers, appears worthy of more thorough evaluation by specific disciplines, breadth must not be sacrificed for depth. For example, we do not want to discover a serious environmental drawback of a new method only after it has been widely recommended. In this case, it might be valid to choose the research team mainly by the principle of "making sure we have all the right disciplines."

Discovery of new principles. Some phenomena remain unexplained despite several disciplines' combined efforts. Conceivably, the question requires a new kind of analysis: a disciplinary synthesis. To discover principles beyond those of the contributing disciplines, an unusual kind of scientist is needed: someone who has a thorough command of the analytical techniques and concepts of a discipline, but whose scientific imagination has not been hemmed in by disciplinary training. Here, putting various specialties to work on the

57

same question is least likely to be fruitful. Willingness to learn the language or even the substance of other disciplines helps, but it is not enough: the elusive quality called scientific insight is needed.

Assuming the appropriate scientific talent has been enlisted, it must be used differently than in the kinds of work previously discussed. The individual project—one group of people, working together for a few years—is not the level at which to expect measurable progress. The goal requires a long-term commitment, and we should not expect individual projects to produce self-contained "deliverables" in the sense of other research modes.

The Relevance of Multidisciplinary Research
to Alternative Agriculture

Discussions of research on alternative agriculture usually list "multidisciplinary" or a similar term among the leading characteristics (Gliessman, 1987; National Research Council, 1989b, p. 14; MacRae et al., 1989). Some program descriptions and granting agency guidelines make it almost a matter of definition that alternative agriculture research is multidisciplinary (e.g., Madden and O'Connell, 1989). But "alternative agriculture" encompasses various ideas, and "multidisciplinary" comes in various forms, each suited to different research goals. The connection between the two thus cannot be as simple as it is sometimes made out to be.

SPECIAL CHARACTERISTICS OF ALTERNATIVE SYSTEMS

Why might alternative agriculture, especially, demand a multidisciplinary approach? One reason is circumstantial: disciplinary boundaries, in part, reflect the agricultural system we now have, whereas alternative systems, by definition, diverge from common practice. Thus new research on chemical control of weeds, for example, will easily find a clear disciplinary home—weed science—because chemical weed control already is important and has strongly shaped the development of weed science as a discipline. In contrast, a strategy that integrates weed control with control of diseases and insect pests is not likely to fit into one department because that approach is not common, and arose after existing disciplines were established.

58

Other grounds for expecting alternative agriculture research to be multi-disciplinary have to do with its substance, not the fact that it is alternative. Three possible reasons, from the most convincing to the most speculative, are the following:

• A proper evaluation of an alternative system should cover all its relevant aspects.

• Farmers have been an important source of alternative agriculture innovations.

• An understanding of agroecological principles is essential in designing alternative systems.

All-inclusive evaluations. Interest in alternative agriculture is a response to the belief that conventional research has not given enough attention to several important consequences of our agricultural system: environmental effects, implications for rural communities, depletion of nonrenewable resources, and long-term changes in soil productivity. Therefore, one can hardly ignore these considerations when evaluating alternative systems. This argues for a multidisciplinary approach; specifically, the "additive" mode may be adequate.

Farmers' innovations. Some alternative systems, notably organic farming, were used by farmers before they were studied formally. As we have explained, a nondisciplinary mode of research can help insure that promising farmer innovations are not overlooked.

Agroecological principles. Alternative systems are commonly said to be more diversified and more dependent on interactions among their components. The interactions are thought to explain why such systems can control pests and supply plant nutrients without the chemicals needed for these tasks in simpler conventional systems (Francis and King, 1988).

The development of systems based on this principle requires a thorough understanding of how production techniques are coupled to each other and to the environment (Gliessman, 1987, 1990b). Moreover, the components that

59

must be considered are as disparate as crops, livestock, insects, pathogens, soil invertebrates, and humans. Such a goal calls for the synthetic kind of multidisciplinarity described earlier. It would require an innovative, long-term merger of the perspectives of ecology with those of agronomy and other agricultural sciences (Paul and Robertson, 1989; Elliott and Cole, 1989). In the next chapter, we describe efforts to achieve this.

"NONSPECIAL" CHARACTERISTICS OF ALTERNATIVE SYSTEMS
These arguments do not mean that "alternative" necessarily equals "multidisciplinary." A single-discipline approach can play a major role even in research with an alternative orientation. On the other hand, these arguments may have even greater significance than is currently claimed because they apply to more than alternative systems.

We should take account of all important effects of any system, no matter who advocates it or why. Similarly, even though farmers' innovations in alternative agriculture have received prominent publicity, farmers of all orientations come up with many kinds of innovations that deserve to be examined with an open mind. Finally, the claim that alternative systems are especially dependent on system-level agroecological principles is not obvious. We consider this point further in chapters 6 and 7.

Thus, the characteristics that make multidisciplinary research appropriate to alternative systems need not confine it to such systems. We expect that both multidisciplinary and single-discipline approaches will continue to be important for a wide spectrum of projects, whether "conventional" or "alternative."

Specialization and Fragmentation: A Failure of Institutions or an Inevitable Scientific Tendency?

Is agricultural research as a whole, beyond the level of the individual project, too specialized and not sufficiently multidisciplinary? If so, the reasons could be institutional: the reward system, funding mechanisms, the departmental structure of universities, the channels of professional communication, and so forth (Campbell, 1969; MacRae et al., 1989). If these are the reasons, the solution is to reform the profession.

On the other hand, some critics believe that such reforms would not be enough because of a deeper problem: the way scientists view the world. For example, Busch and Lacy (1983, pp. 240–42), reporting their study of the social and organizational aspects of agricultural research, first considered institutional changes to improve the research system. They then turned to "epistemological considerations," believing that "institutional changes . . . are probably insufficient to surmount the problems facing . . . the agricultural research community today." For them, the reason was "the unidimensional understanding of the nature of knowledge that appears all too frequently in current research." Similarly, according to one supporter, agroecology is valuable because it "challenges the Western concepts of objective knowledge" (Norgaard, 1983, p. 7).

In the view of these critics, scientific thinking is based on "reductionism," which divides complex phenomena into ever simpler components, obtaining the larger picture simply by reassembling the pieces. A reductionist analysis does not entail any higher-level integration when it recombines the components into a system; instead, it treats the whole as the sum of its parts. In contrast, agroecology has been described as "holistic while conventional sciences are atomistic" (Norgaard, 1983, p. 9).

In examining this criticism, we need to concentrate on the substance of research, temporarily disregarding customary disciplinary labels, which in part reflect institutional conventions and historical circumstances. There is a difference between a discipline as an intellectual area and as a professional structure. According to Ben-David (1971, pp. 143–45), the defining characteristic of a discipline is that the choice of topics is determined by the internal state of the field; the goal is to advance disciplinary understanding as such. For the divisions of applied or "problem-oriented" sciences, including agriculture, he preferred the term "quasi-disciplines." However, most people refer to these divisions as disciplines too, because they have similar professional functions and because both provide for information exchange among scientists and establish standards for rewarding achievements.

If we do not assume that the institutional divisions of science necessarily correspond to divisions in scientific thought, there is less support for the view that science inevitably becomes more specialized and fragmented. We have two grounds for this assertion. First, proliferation of disciplines does not come about only by fragmentation of existing disciplines. Second, in seeking to understand complex systems, multidisciplinarity is not the only path: disciplinary boundaries need not preclude the integrated thinking that is needed.

How New Disciplines Arise: Broadening and Fragmentation

The proliferation of disciplines appears to show that scientific research is highly fragmented and specialized. But when we examine the ways that new disciplines arise, this assumption seems less justified.

FROM SYNTHETIC MULTIDISCIPLINARITY

We have already discussed one way: synthesis of a new discipline from several disciplinary viewpoints. When the number of disciplines increases this way, the result is not fragmentation of science, rather the expansion of the ideas it encompasses.

Agroecology is an attempt at such a synthesis. If successful, it will build on an earlier synthesis that established the field of ecology. However, despite ecology's initially synthetic tendency, the divisions between terrestrial and aquatic and between plant and animal ecology have persisted for decades (McIntosh, 1976). In other words, disciplinary development can be a force for both synthesizing and fragmenting.

FROM EXTENSIVE MULTIDISCIPLINARITY

A new discipline may arise after people from several disciplines have focused on a topic of common interest long enough and with enough success that they decide to institutionalize such collaborations. Dairy science developed this way. For a long time, dairy cows had been studied by researchers in several disciplines, such as physiology, nutrition, and genetics. Working separately, they could not supply the greater knowledge demanded when dairying became a modern, highly technical industry (Rossiter, 1979). However, collaboration gave them a recognized field with its own journals, departments, and scientific meetings. They no longer are scientists who study dairying; they are "dairy scientists." This new discipline is an example of fragmentation. However, it came only after a change in the opposite direction that joined people whose training was different but who were interested in the same animal.

BY SEPARATING FROM A PARENT DISCIPLINE

Separating from a general discipline seems like an obvious case of fragmentation and specialization. Yet even though the result is an increase in the number of disciplines, separation has aspects of the opposite process. In some ways, it exemplifies for a whole discipline the integrated type of multidisciplinarity described earlier: two disciplines combining to study the same topics more thoroughly.

Agricultural economics, for example, was established early in this century to deal with the broad field of farm management. This field required people who not only knew economics but who also could draw on several production-oriented fields, such as agronomy and animal science (Dobbs, 1987). Thus agricultural economics represented integration, not fragmentation—at least at first. The emergence of this new field did not isolate its members intellectually. Several of its early achievements were taken up by economics as a whole and contributed to many important ideas: the law of diminishing returns, comparative advantage in international trade, opportunity costs, linear programming techniques, and productivity indexes (National Research Council, 1972, pp. 54–55).

Unfortunately, a field that begins by incorporating material from other fields may become too specialized because it severs itself from its multidisciplinary origins. This can happen if one parent discipline comes to dominate the others. In agricultural economics, it resulted from rapid advances in econometrics and modeling techniques. When applied to agricultural questions, they helped turn agricultural economics into a subdiscipline of economics (Dobbs, 1987). Thus the field no longer required a strong knowledge of both economics *and* agriculture. Instead, an agricultural economist could be an economist who knew enough about agriculture to apply general economic techniques to agricultural questions.

Such a transformation may be carried a step further, with the subdiscipline becoming a discipline in its own right, not just a specialty within a discipline. At some universities, a student can set out to become an agricultural economist instead of an economist with additional knowledge of agriculture. This path may mean forgoing important background areas of general economics (National Research Council, 1972, p. 410). However, this development has not been universal: agricultural economics sometimes remains part of the general economics department.

But under either arrangement, agricultural economists may abandon their field's traditional problem-solving orientation, seeking the higher prestige of more theoretically oriented "pure" economics. (The tendency to emulate one's "purer" disciplinary counterparts occurs in many branches of agricultural research, as we discuss in chapter 11 in connection with the professional reward system.) One agricultural economist has warned that "a badly flawed notion of what agricultural economics is about" is pushing some departments to "surrender the goals and culture of agricultural economics to that of economics," turning them into "at best, second-rate economics departments of which there is already a sufficiency" (Bonnen, 1986, pp. 1078–79).

Thus a field that arose by combining other fields has become another example of fragmentation and specialization, because only portions of both parent fields remain involved. The agricultural content was subordinated to the economics, and the economics became more specialized. Similarly, rural sociology, agricultural entomology, and several other disciplines have largely separated themselves from their parent disciplines. In land-grant universities, the same campus may have one department for an agricultural discipline and another for the corresponding general discipline without its agricultural component.

The parent field often loses by the separation, too. For example, Levins (1973, p. 523) described the tendency of nonagricultural biologists to denigrate agricultural species as "merely applied," whereas species remote from human activity are "fundamental." (Corn, he noted wryly, is an exception, having figured in fundamental research "in spite of being edible.") He called on biologists, especially ecologists, to recognize that "there is a rich intellectual content in the study of agriculturally relevant systems," and that "agriculturally relevant research is also of fundamental significance" (p. 523). This is a key point in our discussion of agroecology (chap. 6).

Is Reductionism the Problem, and Is Multidisciplinarity Its Cure?

REDUCTIONISM AND THE SCIENTIFIC WORLDVIEW

We have tried to show that scientific thinking sometimes is more multidisciplinary and less fragmented in substance than one might think from the proliferation of disciplines. Furthermore, even strictly single-discipline work is not necessarily so reductionist that it cannot deal with complex systems such as alternative agriculture. Multidisciplinarity is not the opposite of reductionism nor the only alternative to it, although it has been called exactly that (Bella and Williamson, 1976).

A related belief in need of examination is that reductionism assumes that "nature, or any reality, is simple and must 'make sense'" (MacRae et al., 1989, p. 179). This supposedly renders it unsuited for alternative agriculture because "complex biological systems do not always make sense" (MacRae et al., 1989, p. 179). Some critics go even further, belittling reductionism as "the prime paradigm for an emphasis on cleverness" (Jackson, 1990, p. 416). What they advocate, therefore, is a more synthetic and more insightful way to deal with complex, diversified systems by combining diverse disciplinary perspectives.

We have already mentioned the extreme version of this view: that what limits scientists' thinking is nothing less than "Western concepts of knowledge." (Presumably, the term "Western" is intended to reflect the narrowness of scientists' cultural perspectives. It therefore is interesting to note what one of "Western" science's undisputed giants has said about it: "For a parallel to the lesson of atomic theory . . . we must in fact turn to . . . epistemological problems with which already thinkers like Buddha and Lao Tse have been confronted, when trying to harmonize our position as spectators and actors in the great drama of existence" [Bohr, 1961, pp. 19–20].)

If the limits of science are so fundamental and universal, we need not restrict the discussion to agriculture. Why not look at other branches of sci-

65

ence, too? Physics is especially suitable as an example because one of its major achievements, Newtonian mechanics, is considered the pinnacle of mechanistic, objective, deductive science. This kind of science, sometimes labeled "reductionist," is said to be what limits our ability to understand alternative agricultural systems:

> [A] fallacy that threatens our chances of making intelligent choices for a sustainable future in agriculture is reductionism. Reductionism is . . . based on 16th and 17th century science, which provided an abstraction of the world that reduced reality to a mechanistic system consisting of two basic entities: matter and motion. That mechanistic view of the world still influences modern perceptions. It is this reductionistic view of the world that causes us to assess the performance of agriculture in narrow, microcosmic units and to judge the economic health of a farm based on an individual farm's year-end bottom line, rather than the decade-long prosperity of groups of farms within rural communities. (Kirschenmann, 1991, p. 167)

BEYOND NEWTON

But what about the past three centuries of science? Post-Newtonian developments refute two related claims: that Western scientific thought necessarily is reductionist, and that reductionism cannot deal with phenomena that defy common sense (Lockeretz, 1991b).

Thermodynamics and its more abstract offshoot, statistical mechanics, represent the exact opposite of reductionism: the characteristic concepts of both, such as entropy, are defined only for systems with a great many constituents. These concepts let us understand the behavior of such systems in a way that could not have been deduced from the individual behavior of their constituents. They also show that systems with many constituents share certain properties, no matter how different their constituents are. Thus they are "system-level" or "emergent" properties, like those that make an ecosystem "different than the sum of the parts" (Jackson, 1990, p. 416). Supposedly, prevailing scientific thought cannot deal with such properties because it is too constrained by reductionism.

66

That "reductionism" cannot cope with "counterintuitiveness" is disproved by quantum mechanics, in which the fundamental idea is that ordinary reasoning and day-to-day experience do not apply to submicroscopic phenomena. For example, in quantum mechanical theory, a particle that goes through a barrier with two openings need not go through either one or the other. By their analysis of such questions in the 1920s, Bohr, Heisenberg, Schrödinger, and others forced a profound reexamination of established notions like causality and observability.

The result was a radical change in the most basic scientific ideas: how the measurement process interacts with the object being measured; how the choice of terms to describe a phenomenon affects the theories we develop about it; and how and why a scientific theory must take account of the observer. In short, such work shows that scientific reality consists of considerably more than just "matter and motion."

Finally, the general theory of relativity is both strongly counterintuitive and nonreductionist. It is hard to imagine a less reductionist scientific development than an explanation of the dynamics of the entire cosmos. The explanation required some remarkably counterintuitive ideas: that ordinary space and time are part of a curved four-dimensional space, and that the mass of an object is affected by the distribution of all masses in the universe. General relativity foreshadowed some alternative agriculture precepts by half a century—for example, the statement that "according to ecological thinking, nothing is completely self-contained. Everything is what it is by virtue of its relations to other things" (Cobb, 1984, p. 211).

HOW LIMITING IS SCIENTIFIC THOUGHT?

These examples make it hard to take seriously any discussion of agricultural research that assumes that scientists' ideas about the nature of knowledge have not gone anywhere since Newton. They also refute the assertion that scientists need new modes of thought to deal with system-level and counterintuitive phenomena. Most of all, they show how wrong one can be in asserting the limits of "scientific thinking" on the basis of a single field, such as one's own.

Understanding what scientific thinking can and cannot encompass is not just a matter of abstract philosophy. We saw its practical implications in the earlier quotation that purported to link farm problems to the "fallacy of reductionism." Similarly, scientists' outlook is said to be crucial in determining the kind of agricultural system we will have: "A shift of thinking from that which is based on a mechanical model to that which is based on an ecological model . . . is necessary if we are to have a sustainable agriculture" (Cobb, 1984, p. 210).

We believe that science has already broken through some barriers that are said to prevent it from handling alternative agriculture. In making this point, we did not have to draw our examples from physics, although, as we explained earlier, physics is the science least likely to appeal to critics of reductionism and a mechanistic world view. Population genetics also is "Western" science. It simultaneously deals with a large-scale biological phenomenon and the smallest component of living organisms, the DNA sequence. The "Arcadian" natural historians of the late eighteenth century also were "Western" scientists (Worster, 1985, chap. 1). Their interest in the complex interactions among ecosystem components was even more in keeping with what many advocates of alternative agriculture are calling for today (McIntosh, 1985, chap. 1). Very much a "Western" scientist, too, was the USDA's chief chemist, who made the following comments several decades before the concept of organic farming was formulated:

> The more thought I give to the question the more I am convinced it is true that the soil is a living organism. . . . When a man sends to me a specimen of soil and writes, "Please analyze this soil and tell me what crops I can grow on it," I send him word, "Ask your soil itself what you can grow on it; by asking your question directly to the soil, you can get a better answer than in any other way." (Wiley, 1904, pp. 142–43)

ARE THERE LESSONS FOR AGRICULTURE?

Does this discussion have anything to do with agriculture? We did not digress into physics because we think that the agricultural sciences or any others should take physics as their model. On the contrary, we strongly agree with Ben-David (1973) on why the social sciences, for example, should *not* do so. Nevertheless, the examples from physics are informative.

First, success in physics did not come from an exclusively reductionist approach, but neither was reductionism discarded. In contrast, proponents of alternative agriculture may reject reductionism outright, although some agroecologists emphasize the importance of working simultaneously at several hierarchical levels of the ecosystem (e.g., Hendrix, 1987).

Second, success arose within physics itself, when it tackled previously unknown phenomena and extended its thinking accordingly. It did not have to bring in another discipline to be able to handle more complex systems. Thus, even if excessive reductionism is the reason a discipline cannot deal with complex systems—and this diagnosis requires a more careful examination than it often gets—multidisciplinarity is not always the cure. The fact that science is organized by disciplines does not automatically mean that its thinking is narrow and inflexible. An inadequate range of disciplines in a research project is easier to fix than an inadequate scientific imagination. But especially in research that hopes to achieve a true conceptual breakthrough by synthesizing different perspectives, it is not usually the main limitation.

6

· · ·

The Potential Contribution of Agroecology

Modern science, favored as never before with the means of extension and development, should be able to justify its cost to the state by contributing to the betterment of human life. Tested by its capacity to meet this demand, ecology, I think, will not be found wanting. Agriculture, horticulture and forestry are, consciously or not, practical applications of its principles, and their best development has been attained where these principles have been most intelligently observed and applied.—Spalding, 1903, p. 209

Proposals for change in agricultural research, especially those oriented toward alternative agriculture, commonly call for greater attention to agroecology, agroecosystems, and the "systems thinking" associated with ecology. Agroecology (or agricultural ecology) is the merger of analytical methods, concepts, and data from ecology with more traditional approaches for studying agricultural problems.

The merger can benefit both fields. Agricultural scientists can gain insights into how agroecosystems function; this understanding in turn can help them develop production systems that depend less on agrichemicals, use natural resources more efficiently, and cause less environmental damage. On the other side, ecologists can benefit by drawing on the wealth of agricultural data and analysis that has been accumulated. Also, they can study ecological processes under conditions that are more controlled than in natural systems.

The merger poses difficulties, however, especially the temptation to try to do too much. Agroecology can help analyze agricultural systems in greater breadth. But it is not adequate for studying the social aspects of agriculture, as some of its advocates claim.

In this chapter we try to delimit how agroecology might improve our agricultural system. This entails the following questions:

• How is agroecology related to the concept of alternative agriculture?

• What are the scope and goals of the principal versions of agroecology, and why are the boundaries of the field so unclear?

• What has come out of previous attempts to merge ecology and agricultural science, and why has the merger been so troublesome?

• Can ecology deal effectively with social phenomena?

• How can agroecology contribute to developing sound agricultural policy?

The Connection between Agroecology and Alternative Agriculture
Agronomists and ecologists have suggested more interaction between their fields for several decades. Since the 1980s, the plea for greater collaboration has been heard especially frequently in alternative or sustainable agriculture. Many people see it as a solution to the adverse environmental and social consequences of modern agricultural technology. Two of the first scientists to make this point introduced their book *Agricultural Ecology* with the challenge to "apply an ecologically sound agricultural technology to increase production without, in the long run, destroying agricultural lands or damaging global ecology" (Cox and Atkins, 1979, p. 5).

The connection with sustainability has been asserted by many people:

• Agroecology "establishes a framework for long-term sustainability of agricultural systems" (Gliessman, 1984, p. 161).

• "Sustainability requires new directions for agricultural development, directions based on the principles and practical knowledge of ecology" (Dover and Talbot, 1987, p. 7).

• Agroecology is "the science that can provide the knowledge required to achieve agricultural sustainability" (Gerber, 1990).

Dover and Talbot (1987, pp. 21–28) went beyond asserting a connection, offering an explanation of it. They suggested that among the ecosystem properties studied by ecologists, stability and diversity are especially relevant to understanding and improving agroecosystems. The meanings of these two concepts and their interrelationship have been analyzed more deeply by ecologists than by agricultural scientists. Sustainable agriculture is necessarily stable in the sense of being able to resist or recover from environmental and economic fluctuations. Therefore, ecologists' understanding of stability can make an important contribution. Some people have gone further, suggesting that stability of a farm requires diversity of crops and livestock. However, ecological theory has not clearly shown a causal connection between species diversity and the ability to withstand environmental perturbations. Nor does ability to recover from environmental perturbations insure economic stability.

The Boundaries of Agroecology

"HARD" AND "SOFT" AGROECOLOGY

One difficulty in understanding what agroecology can contribute to agriculture arises because it comes in several versions. They share the idea of addressing agriculture's environmental problems, but they differ in other goals.

One version (e.g., Loucks, 1977; Lowrance et al., 1984; Elliott and Cole, 1989), sometimes called "agroecosystems science," might be termed "hard" agroecology. Its proponents recognize that agroecosystems are dominated by human activities, but they focus on its nonhuman side. That is, they deal with phenomena traditionally studied in systems ecology, such as energy, material flows, and nutrient cycling. They use simulation modeling and sophisticated software, such as geographic information systems, to predict the consequences of agricultural management options. Occasionally, modeling and other techniques seem to take precedence over the scientific purposes or social utility of agroecology; social purposes are not the driving force behind "hard" agroecology.

In "soft" agroecology, in contrast, social and political aims are fundamental. In this view, agroecology is important for restoring or maintaining a more "natural" agriculture (Soule and Piper, 1992, pp. 124–45). Its ideas for environmentally sound farming methods draw heavily on "nature," such as "the few relics of pre-Columbian vegetation that remain" (Jackson and Piper, 1989, p. 1591). It also focuses on traditional agricultural systems as research sites and as models for designing farming systems. "Soft" agroecology does not clearly separate the social from the biological aspects of farming systems. On the contrary, combining them is a stated goal: "The interdisciplinary perspective of agroecology also encompasses the field of cultural ecology" (Gliessman, 1984, p. 170).

This version of agroecology is sometimes promoted as a strategy for agricultural development, serving "as a focus for the training, research and information networking activities . . . to further the development of peasant agriculture" (Altieri, 1990, p. 118). It also is relevant to agriculture in industrialized societies (Altieri, 1983; Jackson and Piper, 1989).

"Soft" agroecology resembles the analytical method known as agroecosystem analysis for research and development, developed by Conway (1986), which is based on agricultural ecology and human ecology. This approach is a response to many of the same environmental and social problems of agricultural development that worry "soft" agroecologists. The difference is that Conway recognized that agroecology does not generate research questions and solutions independently. Rather, after a preliminary multidisciplinary analysis of social problems has identified research questions, agroecology can be one of several tools to address them. Conway's scheme is an example of an appropriate use of agroecology. In contrast, inappropriate expectations of agroecology are particularly common among "soft" agroecologists.

INTERMEDIATE VERSIONS OF AGROECOLOGY

Other versions of agroecology fall between the "hard" and "soft" extremes. Like the "hard" version, intermediate positions emphasize scientific goals that most ecologists and agronomists hold in common: understanding how agricultural systems function and improving their environmental and agronomic performance.

However, as with "soft" agroecology, proponents of intermediate versions regard agroecology as especially relevant to sociopolitical issues. For

73

example: "Agroecosystems may be judged according to the goods and services produced, their contribution to human needs or happiness, and their relative distribution among the human population" (Gerber, 1989, p. 36). Dover and Talbot (1987, p. 29) emphasize the role of ecology in achieving desirable sociopolitical goals, suggesting that planning for agricultural development should be based on ecological principles to prevent resource depletion and environmental degradation. But unlike the "soft" version, intermediate agroecology does not see social and political change as the main goal, and its proponents do not assume that it necessarily will contribute toward achieving such change.

RADICAL AGROECOLOGY

Despite their differing purposes, "hard," "soft," and intermediate versions of agroecology all rely on standard scientific concepts. The term agroecology also is used by some people to mean a radically different kind of science. For example:

Agroecology is a vanguard movement in opposition to the atomism, reductionism, and materialism of the prevailing but obsolete paradigm. . . . [It] provides a systemic, integrative, holistic understanding of reality at the level of organic terrestrial nature. (Callicott, 1988, p. 6)

Despite assertions that "agroecology stems from different roots than those of most Western science" (Norgaard, 1983, p. 7), ecology is firmly rooted in the same intellectual traditions as other natural sciences and agriculture (Egerton, 1976; McIntosh, 1985, chap. 1). Thus agroecology does not automatically acquire a radical purpose simply because it fuses ecology with agricultural sciences. However, radical agroecology's sharp divergence from other scientific schools of thought has colored the way many people near the "soft" end of the spectrum view agroecology.

THE BOUNDARIES OF AGROECOLOGY IN PRACTICE

Given these wide differences in agroecology as a concept, another way to learn its scope is to look at what practicing agroecologists study. Edited or

74

multiauthor books are a reasonable place to get an overview of the field. Prominent recent examples are Cox and Atkins (1979); Lowrance et al. (1984); Gliessman (1990a); and Carroll et al. (1990). These books reflect the main reasons for the contemporary interest in agroecology, especially the quest for environmentally sound farming methods.

Unfortunately, these books generally lack coherence and are poorly focused. Their coverage ranges from broad discussions of climatic processes and pedogenesis to the politics of hybrid corn. Other surveys of agroecology, such as those of Azzi (1956) and Tivy (1990), are less eclectic and more narrowly focused. However, these versions of agroecology are not the ones that recently have been receiving great attention in connection with agriculture's environmental problems.

Why Agroecology Is So Diffuse

The tendency to encompass too much—particularly the attempt to include social analysis—is a serious problem that must be resolved before ecology can make significant contributions to agricultural science. It is sobering that ecology's lack of recognition in the agricultural curricula of land-grant universities "is almost as true today as it was in the last century" (Paul and Robertson, 1989, p. 1595). If the merger were more clearly limited to what is feasible, perhaps ecology could give important support to agriculture.

Why do many agroecologists have such amorphous and overly ambitious goals? The problem springs in part from several inaccurate notions: what ecology covers; the degree to which its perspective is holistic; and its relation to political issues.

Frequently, agroecology's advocates think of ecology as concerned mainly with preserving natural resources, so that incorporating its perspectives into agriculture necessarily will advance that goal. But practicing ecologists have other goals too, such as improving how resources are managed for human consumption (Sagoff, 1985). Alternatively, when they test hypotheses about how species, populations, or systems function, they may have no purpose other than advancing ecological theory.

Another reason for the diffuseness is the common belief that ecology and therefore agroecology necessarily are holistic. For example: "Through the application of ecological concepts and principles to the design and management of agricultural systems, a holistic perspective is established" (Gliessman, 1990b, p. 367). The assumption that holism is the basis of agroecology's distinctiveness is strong among purveyors of "radical agroecology" (e.g., Norgaard, 1983, p. 10). However, much ecology is not holistic at all; ecologists often focus on single organisms or limited relationships.

A third reason for inflated expectations is an imputed political stance that is not inherent in ecology. For example:

> The agroecological approach focuses on the system as a whole with an eye toward the multiple goals of productivity, profitability, reduced uncertainty and vulnerability, equity, protection of the health of producers and consumers, environmental protection, and long-term sustainability and flexibility. (Levins and Vandermeer, 1990, p. 342)

Some of these goals, such as productivity, are amenable to scientific analysis; others, such as equity, are matters of public policy. Political purposes may be no less worthy of attention than improving scientific understanding, but they will not be achieved simply by combining ecology with other physical or biological sciences.

Similarly, Jackson and Piper (1989, p. 1593) consider that a "properly matured and seasoned" ecology can address most problems that have beset industrialized agriculture, including social, political, economic and even religious issues. But even if it could do so, the only reason would be that the "seasoning" process had created something very different from what ecology is now. True, many ecologists are interested in sustainable agriculture, conservation, and human effects on the physical environment. However, these are not strong research areas in mainstream ecology; for example, they are not among the topics studied in the most frequently cited ecological papers (McIntosh, 1989).

Misplaced expectations of ecology also can come from scientists' sincere commitment to political change and their desire to integrate their political principles with their scientific work. This tendency may be reinforced by public pressure to justify science by its social relevance. Some ecologists find their work meaningful because it is connected with both environmental problems and social causes. Emphasizing the latter can cause biological and physical scientists to forget the limitations of their disciplines. Those who have had little contact with social sciences in their training often do not understand that social problems cannot be analyzed simply by extending the methods used to study ecological problems. We return to this point after considering how ecology can be used to analyze the biological side of agricultural systems.

Joining Ecology with Agricultural Sciences

PREVIOUS ATTEMPTS AT MULTIDISCIPLINARY COLLABORATION
The need for closer collaboration between agricultural scientists and ecologists has been a recurring theme in both fields since the turn of the century. Significant ecological theory has been based on the study of agricultural pests and crops, and many concepts developed by ecologists who studied natural ecosystems are applicable also to human-managed systems. One early plant physiologist believed that ecology would "undoubtedly hold a commanding position in the curriculum of the agricultural and general science courses" of agricultural colleges (Arthur, 1895, p. 368). Nevertheless, ecologists and agricultural scientists have followed diverging paths through this century, largely ignoring each other (McIntosh, 1985).

Disciplines such as entomology and microbial ecology benefited from early cooperation between ecologists and agricultural scientists. Several early ecologists worked at agricultural colleges, including John Weaver, F. E. Clements, J. T. Curtis, and Aldo Leopold. However, agricultural scientists and ecologists at agricultural colleges did not develop strong, long-lasting ties (McIntosh, 1985, p. 34). Current definitions of agroecology emphasize interdisciplinarity (Cox and Atkins, 1979, Preface; Gliessman, 1984), but most work that has contributed to agroecosystem analysis has been at the disciplinary or subsystem level. Integration at the ecosystem level or among disciplines has been less common (Loucks, 1977).

DIFFICULTIES FACING A SYNTHESIS

Productive integration of ecology and agricultural sciences is impeded by the same problems as other multidisciplinary efforts, plus additional obstacles specific to these fields. Some of the obstacles are institutional, including the orientation of state land-grant universities toward the immediate solution of local problems rather than a broader ecological perspective (Buttel and Gertler, 1982), and the reward system of the profession, which inhibits agricultural researchers from pursuing unconventional approaches.

Another barrier is the kind of question that each field favors. Ecologists have preferred problems of theoretical interest, while agricultural scientists have studied problems with more practical applications (Cox and Atkins, 1979, Preface). As a result, "ecologists tend to view agronomists as strict empiricists, and agronomists tend to view ecologists as overly theoretical purveyors of the obvious" (Paul and Robertson, 1989, p. 1594). Although many agricultural scientists have adopted a mode that resembles the "basic" science style of ecology, their work is supposed to be valuable because it helps solve problems in the outside world, not within their disciplines. A point in common between agricultural scientists and ecologists is that both generally have considered the social sciences to be inferior (as discussed, for example, by Rhoades et al. [1986]).

When ecologists and agricultural scientists work on similar topics, their analyses usually focus on different questions. Even when ecologists study a single organism or population, they relate it to its environment and other controlling factors. Agronomists, on the other hand, until recently have not addressed the complex interactions affecting crop productivity: "Added fertilizers have obviated the need to understand soil organic matter dynamics or microbial community changes in cultivated soils; pesticide availability has often made knowledge about weed and insect life-history strategies or differential competition redundant" (Paul and Robertson, 1989, p. 1594).

Such statements aside, synthetic pesticides and fertilizers cannot be used efficiently without a good understanding of how they affect crops. However, their availability has strongly influenced the topics that many agricultural scientists have chosen to investigate. Weed science and much of integrated pest management now emphasize the effective use of synthetic chemicals. They are only beginning to deal with the full range of management methods that would be possible with a better understanding of pest ecology and physiology. Ecologists, in contrast, study organisms primarily in nonmanaged ecosystems. When they investigate synthetic chemicals, often it is to learn how they alter a natural ecosystem, not how they can be used to manage pests.

Attempts by Ecologists to Deal with Social Phenomena

HUMAN INTERVENTION IN AGROECOSYSTEMS

Analyses that apply ecological methods to natural systems in which humans intervene have dealt with both the effects of human intervention and its causes. Human intervention is the main difference between agricultural and natural ecosystems, and mismanagement is the main reason for environmental degradation. This is true even for agroecosystems that closely resemble natural ecosystems.

When agroecology addresses not only the effects of human intervention in natural systems but also the causes, it veers into domains traditionally studied by social scientists. Many agroecologists espouse the view that their field should include social analysis. For Cox and Atkins (1979, Preface), agroecology "deals with processes in nature, yet relates these processes to man in intimate fashion; it concentrates on scientific phenomena, yet recognizes the ties that exist to economics, politics, and other fields of human activity." Gliessman (1984) expanded this idea, saying that the field should be "more than just ecology applied to agriculture" (p. 170). To him, the development of agroecosystems is a coevolution between culture and environment. Therefore, agroecology must "involve the integration of ecological and cultural knowledge" (p. 171).

Agroecologists differ in how closely they think ecological and cultural knowledge should be combined, and it remains unclear how this merger will happen. Sometimes an ecological study merely concludes with a gesture toward social analysis: "The need to embark on large scale pilot studies of crop diversification should not continue to be neglected. . . . A socio-economic perspective should parallel the biotechnical investigations" (Al-

79

tieri et al., 1990, p. 80). Other agroecologists assume the linkage is stronger, occurring through multidisciplinary teams made up of social, agricultural, and natural scientists.

CONTROVERSIES OVER APPLYING ECOLOGY TO HUMAN AFFAIRS
Previous attempts to apply biological or ecological principles to humans have been strongly criticized. Agroecologists who advocate the integration of social and ecological studies in agroecology should first understand why similar efforts have proven so difficult.

The most prominent attempt at a synthesis of social and biological sciences was Wilson's *Sociobiology* (1975). This book brought forth a flood of criticism from social scientists, biologists, and ecologists (e.g., Caplan, 1978; Montagu, 1980; Sahlins, 1976). Among the arguments against sociobiology were the paucity of direct evidence for the presumed genetic control of human behavior; the inadequacy of its "just-so" stories as the explanation of human adaptation; its implicit toleration of existing social inequalities as natural and unavoidable; and its reduction of individual human achievements to instinctual survival strategies. Wilson offended many social scientists by trying to subsume social science within biology: "[Sociobiology] has sought to be the final arbiter on many areas within the human disciplines, when at best it can be an informative adjunct, paralleling and enlightening traditional approaches" (Heyer, 1982, p. 226).

More recently, H. T. Odum has led attempts to merge the study of ecological and human topics in various systems, including agriculture, by using models that combine physical and biological variables with cultural components such as family and religious institutions (Odum, 1983, chap. 24). Although thought-provoking, these efforts are subject to some of the same criticisms as sociobiology. In particular, by treating social concepts in the same way as material and energy flows, the models overlook the uniquely human characteristics that make these concepts so complex: "In fact, in feedback systems of energy or information the individual is potentially obsolete—all processes, whether physical or biological, are substitutable" (Taylor, 1988, p. 244).

Forced syntheses of biology or ecology with social sciences can lead to unrealistic hypotheses and unpalatable policy prescriptions. Problems that require sociopolitical analysis and solutions should not be squeezed into analytical schema appropriate only for inanimate objects and nonhuman life. Doing so creates a danger parallel to the technocratic viewpoint that many agroecologists criticize in modern agricultural development.

Attempts by Social Scientists to Deal with Ecological Phenomena
THE UNSATISFACTORY SYNTHESIS OF ECOLOGY AND SOCIAL SCIENCES

Social scientists have tried to incorporate ecological concepts more often than ecologists or agricultural scientists have tried to deal with social phenomena. An early attempt by anthropologists and geographers was environmental determinism: the idea that human social and cultural behavior is largely determined by the natural habitat. This view lost favor because of its rigid and ethnocentric claims (Ellen, 1982, chap. 1). Social scientists today are wary of attempts to explain social phenomena merely by physical conditions.

Partly in reaction to the extreme stance of environmental determinism, the most influential schools of thought in social sciences in this century "assume that various *human* 'mechanisms'—social institutions, culture, technology, and so forth—will operate to insure that a human population will adapt successfully to its biophysical environment" (Dunlap, 1980, p. 7; emphasis in original). According to this "human exemptionalist paradigm" (Catton and Dunlap, 1980), humans are immune to physical factors that influence the distribution and abundance of other organisms, such as vulnerability to scarce resources.

Because this view has been so dominant, social scientists who have tried to deal with the physical environment and the organic basis of human behavior have historically been out of the mainstream (van den Berghe, 1978). Only recently has there been general acknowledgment that human societies depend on limited physical resources, but how these limits will affect human societies remains an open question.

"Human ecology," an attempt to join ecological concepts and principles with social science, has been re-created several times. However, there is little agreement about what it is (McIntosh, 1985, p. 315). Human ecologists use concepts and terms familiar from ecology, such as competition, natural selection, and material and energy flows. However, such extrapolations to the dynamics of human societies do not usually hold up to critical examination. They can stimulate insights into the mechanisms that control human systems, just as metaphors can enrich communication. But if they are taken too literally, they obstruct social analysis with inappropriate jargon and lead to a search for nonexistent correlates between human societies and natural ecosystems.

A related problem that social scientists have had in incorporating ecological concepts is their superficial understanding of what these concepts mean to ecologists. This superficiality parallels ecologists' and agricultural scientists' inadequate understanding of social science concepts. The unfortunate result is that natural and agricultural scientists become impatient with social scientists' writings on agriculture or ecology because the facts and the science often are wrong. This also can be a problem with courses on the social aspects of agriculture offered to agricultural science students. Because the two areas have a long-standing lack of respect and collaboration, few professors can combine them satisfactorily.

DIFFERENCES BETWEEN BIOLOGICAL AND SOCIAL SCIENCE ANALYSES

Most efforts to merge social sciences with ecology have been deficient for the same reason: humans and nonhuman organisms require different forms of analysis and explanation. The fundamental problem of the "ecological perspective" in anthropology has been framed this way: "Is there a cultural equivalent of Darwinian fitness that controls the rate at which cultural variants are passed on?" (Hardesty, 1980, p. 113).

No such general theory has been devised. Functional explanations in the social sciences therefore can only cover particular cases. For most social phenomena, one can find some beneficial consequence that "explains" their existence. Such reasoning allows accidental correlations to be mistaken for causes. Despite its flaws, the search for functional explanations of human behavior is compelling; many social scientists assume that all social and psychological phenomena have some meaning that explains why they arise and persist (Elster, 1983, p. 55).

Human society is qualitatively more complex than "societies" of most other organisms because people can act intentionally or irrationally. For example, humans can use indirect strategies, sacrifice their needs for others, delay gratification, persuade others to work collectively, and act self-destructively. In other words, humans as individuals and in groups, do not always act the way the Darwinian model of natural selection predicts.

Our criticisms of functional explanations do not imply that human behavior cannot be explained. Social scientists have developed methods to address questions involving uniquely human behavior. For instance, social phenomena often can be explained by showing how mechanisms such as socialization cause people to hold particular desires that lead to particular intentions. These in turn cause particular effects when the individuals holding them interact (Elster, 1983, pp. 84–85).

The methods of social analysis often lack the precision and control possible in biology. For example, we can describe precisely how the diversity of wildlife species changes when a wetland is drained and plowed. In contrast, we cannot describe precisely how a farmer decides whether to plow a wetland or preserve it for wildlife. The lack of precision usually reflects the phenomenon itself, not a failure of the methods. Ecological methods sometimes seem to yield precise descriptions of both the human and the biological aspects of agroecosystems, but they do so only when they leave out critical human elements.

What Ecology Can Contribute to Agriculture

Collaboration between ecologists and agricultural scientists can take various forms, from multidisciplinary studies in which both fields focus on the same phenomenon, to the synthesis of a new discipline. If the problems mentioned earlier can be overcome, such collaborations will have benefits both for the respective disciplines and for society. The main social benefit is the potential for enhancing long-term agricultural productivity and sustainability (Gliessman, 1984; Elliott and Cole, 1989; Gerber, 1990). Ecologists can offer refined modeling and field techniques and an understanding of how natural systems function. They also can provide data on species and habitat diversity in agricultural landscapes, as well as improving our understanding of weed and insect dynamics and biological soil processes (Eijsackers and Quispel, 1988).

These contributions complement agricultural scientists' wealth of data on regional agricultural practices, and their understanding of how specific agricultural systems function. Combining the two groups' efforts can help both to reverse agriculture's harmful environmental effects and to improve farm management.

Including social analysis in agroecology is more difficult. Obviously, agriculture is about more than soil, climate, livestock, and crops. Yet ecology cannot describe and predict human actions as it does other organisms' interactions with their environment. Human effects on agroecosystems can be measured, but agroecological methods cannot tell us why they have occurred or how society might be changed to prevent undesirable effects.

To answer such questions, agriculture would do better to look to the social sciences, not ecology. Agroecology's most meaningful contribution will be an understanding of what happens in the biological and physical environment. This accomplishment will occur sooner if agroecology does not also try to deal with cultural systems.

Unfortunately, the belief that agroecology can contribute to sustainable agriculture has been broadened into the belief that it has a special ability to solve agriculture's social problems. However, agroecology cannot make political recommendations or set moral imperatives. Blurring the boundaries between political and scientific purposes will prevent it from delivering what it can deliver and will put ecologists in positions of moral authority they are poorly suited to fill:

Science is excellent in addressing particulars, but vague and indecisive in recommending general policies for a society to follow. . . .

84

Science is not in any position to find for any part of society, the agricultural or other, a perfectly rational, coherent, comprehensive, objective, permanent design. It has not done that for itself. How could it do it for farmers? (Worster, 1991, p. 281)

Ecology is a scientific discipline concerned with the abundance and distribution of organisms: "As an objective science, ecology is not directly involved with human ethics, morals or behaviour" (Lewin and Polunin, 1990, p. 177). Similarly, "ecology attempts to explain, not prescribe" (Dover and Talbot, 1987, p. 29). If ecologists advocate a particular ethical or philosophical stance, ecology is in danger of becoming merely a point of view (McIntosh, 1985, p. 323).

Although agroecologists have no special prerogative to decide how systems should be managed, they need not remain isolated from real-world social problems. For example, they can evaluate the ecological consequences of policies intended to achieve various social goals, such as a more equitable income distribution in agriculture, better nutrition for rural families, or more stable farm income during difficult economic times. If sustainability is a policy goal, agroecologists' knowledge can help us understand how various agricultural practices will support or impede its achievement. In doing so, however, "agroecologists must become arbiters in the debate on agriculture, not champions of one side" (Dover and Talbot, 1987, p. 56). They also can help society recognize "good" qualities of ecosystems that

are worthy of love and admiration, and which should be preserved because of their place in our natural and evolutionary heritage. These qualities may include the age, diversity, richness, complexity, authenticity, productivity, uniqueness, or other properties. (Sagoff, 1985, p. 102)

This point usually is made regarding natural ecosystems, but it also applies to how agriculture affects vulnerable species and places.

Agroecological research is one tool available for policymaking; it allows policymakers to devise and evaluate a "more fine-grained agriculture, based on a mosaic of varieties, inputs and techniques each fitting a particular ecological, social and economic niche" (Conway, 1986, p. 16). But it cannot achieve these goals by itself:

> Policy barriers . . . —the tax system, the land market, economic insecurity, and export orientation—set broad parameters or micro-economic and macroeconomic conditions which limit the ability of agricultural researchers to identify ecologically sustainable practices which are at the same time privately profitable for farmers. (Buttel and Gertler, 1982, pp. 116–17)

To achieve social or environmental improvements, therefore, the research of agroecologists must be complemented by appropriate policy changes. In the debates over what those policies should be, agroecologists would do well to remember the limits of their field and restrain their contribution accordingly. As one ecologist has warned (Nixon, 1980, p. 510): "It is a bad bargain to trade . . . credibility for political advantage."

7

. . .

Information and Management

Greater intelligence applied to all agricultural work is the need of to-day. The chemist has revealed the laws, hitherto unknown, by which the farmer in co-operation with nature, may secure greatest results at least cost. By the use of his intellect, by skill, by utilizing the information placed at his disposal, by better methods, one man moves forward to success on the farm; while another . . . loses in the game.—Dye, 1899, p. 49

Alternative agriculture often attaches special importance to how farmers can use information and judgment to replace material inputs (Francis and King, 1988). For example: "As pesticides, fertilizers, etc. are reduced, greater knowledge . . . is required for success" (Stinner and House, 1987, p. 146).

Simply doing without chemicals is not usually satisfactory for farmers, and the idea of using information to fill the gap is appealing. Information and management ability do not customarily appear on agricultural production budgets, but they should count as inputs too. Indeed, "some producers consider information their most valuable internal, renewable resource" (Francis, 1990b, p. 67).

If we focus more research on farmers' use of information, we can evaluate available systems more realistically and develop more effective systems that allow farmers to put information to work. Later we will suggest some ways to do this. First, we separate the different roles that information plays in an agricultural system, to show the various opportunities for enhancing its use. We also will compare the role of information in alternative and conventional systems; although the former are commonly described as more "information-intensive," it would be a mistake to ignore any possibilities for improving how all farmers use information.

Information in Developing and Using Production Methods

Calling a production method "information-intensive" can blur several distinctions:

• Who uses information intensively: the people who develop the method, or the farmers who apply it?

• At what point do farmers need information: when deciding whether to adopt the new method, when learning to use it, or each year the method is used?

• What does using information demand of the farmer: additional effort or superior management ability and judgment?

Farmers vary in their inclination to learn about new methods, their ease in mastering them, and the amount of time they can spend. The distinctions we have drawn show that there are varied ways for farmers to use "information-intensive" production methods. Each will appeal to some farmers more than others. We illustrate these distinctions with different reduced-chemical methods of insect pest control.

DEVELOPMENT-STAGE INFORMATION VERSUS
APPLICATION-STAGE INFORMATION

Classical biological control—the introduction of a parasite or predator to control a pest—is "information-intensive" when it is being developed. Typically, the researchers responsible for choosing and evaluating the control agent need a good knowledge of the agroecosystem into which it will be released. On the other hand, the farmer may need no such knowledge to benefit from biological control. Ideally, the control agent establishes itself throughout the region, keeping the pest from becoming a problem ever again; the farmer may not even realize that the introduced organism is now controlling the pest (Sailer, 1979). However, in some versions of biological

88

control, farmers must avoid poisoning the control agent with pesticides aimed at other pests, or they must change their cultural practices to give the control agent an alternative prey when the pest is not abundant.

If investigators do not find a suitable biological control agent, they must widen the search to include reduced-chemical control methods that require a more active role for the farmer. Typically, they take account of several factors: the behavior of the pest; the mechanisms that control it naturally; the conditions that determine how much damage it does; and other pests or beneficial species that might be affected by the control strategy. Again, the understanding and information that go into developing such a system are separate from the information-related tasks it imposes on farmers.

LEARNING ABOUT A NEW METHOD VERSUS USING IT

The information needed by farmers in using a new method can take several forms (Lockeretz, 1991a). Agroecological studies may provide a specific recommendation that farmers can easily follow, such as planting a crop variety that matures before a late-season pest can do much damage. Farmers who understand agroecosystem principles will probably be more inclined to learn about and adopt the new practice, but once they do, they need no further information to use it.

If such a simple solution is not available, a more active strategy is needed, perhaps including a pesticide. The seriousness of many pests varies strongly from year to year, according to the weather, the vigor of the crop, and the density of predators or parasites. This variability means that a pesticide might not be needed in a particular season. The reduced-chemical alternative might be a way for farmers to learn whether they need to spray. Without it, they might use an "insurance" approach, spraying every year to be on the safe side. Unlike switching to an early-maturing variety, this alternative requires action by farmers every year, not simply a one-time change.

WORKING HARDER VERSUS MANAGING BETTER

The yearly effort could take two forms. Perhaps the decision whether to spray or not can be reduced to a simple, definite procedure that any farmer

89

can use if enough time is available. For example, at a specified growth stage of the crop, count the larvae on x plots per acre, using y row-feet per plot; if there are fewer than z larvae per plot, do not spray. With this method, the farmer replaces a material input by work, not by skill or judgment.

The decision to spray might be more difficult. Its monitoring requirements might be so formidable that not all farmers would be equally adept at making the right measurements and keeping adequate records. (The farmer could bring in a professional pest adviser, but such advisers are not available in all areas.) Perhaps the natural control mechanisms are not understood well enough for researchers to develop simple, reliable decision rules. The procedure may depend so strongly on site-specific variables that the farmer's understanding and experience are critical.

Information in Alternative Agriculture

Reduced-chemical alternatives clearly offer opportunities for farmers to apply ecological knowledge. How well these opportunities are being exploited is another matter. Unfortunately, there has been little study of information requirements of the alternative methods that farmers actually use.

Much of what we know comes from studies of organic farmers, who often report that inadequate information is a problem (e.g., Lockeretz and Madden, 1987). However, organic farming is a more drastic alternative than most farmers are considering. Also, organic farmers' problems in part may reflect the small research and extension effort given to the subject so far (Baker and Smith, 1987). The possibility that information requirements may constrain wider adoption of other alternative methods must be examined more thoroughly.

Among research papers published in the late 1980s on reduced-chemical methods, only about one-fifth dealt in any way with information requirements (Lockeretz, 1991c). (For a comparative group of papers on conventional methods, the corresponding figure was only 2 percent!) Even this figure may be higher than for alternative methods that farmers use. Information-intensive production techniques are considered innovative and sophisticated. Therefore they may be overrepresented in research as compared with farmers' practices. As it is, the research literature offers only modest support for the notion that information has an important role in alternative systems.

Management Ability in Alternative Agriculture

In current research on reduced-chemical systems, any explicit consideration of information involves specific tasks that any farmer can do (Lockeretz, 1991c). Usually, researchers give little attention to the operator's management ability. Thus we cannot say what is required for success with an alternative system. Is it just a matter of extra effort, or must the farmer also have a superior understanding of agroecological principles and a good ability to collect and process information?

The question has a long history. Sophisticated farming once meant appreciating the great advances in agricultural chemistry made in the nineteenth century, as seen in the quotation introducing this chapter. This in turn meant recognizing the value of the new commercial inorganic fertilizers and knowing how to use them to meet the precise needs of a specific crop grown on a specific soil. Increased use of commercial fertilizers was part of a more general transformation of U.S. agriculture into a technologically advanced, industrial-style production system. This transformation favored farmers with superior skills and expertise. Previously, someone of average ability had a reasonable chance of succeeding (Danbom, 1979, p. 142).

Nevertheless, reversing the increased use of chemicals does not imply a return to unsophisticated farming; perhaps farmers can cut back on chemicals by becoming more sophisticated managers (Francis, 1990b). Still, the historical record reminds us not to take for granted that low-input alternatives require farmers to become more sophisticated.

THE LINK BETWEEN ALTERNATIVE SYSTEMS
AND MANAGEMENT SOPHISTICATION

The historical record aside, there are reasons that alternative systems indeed might require superior ability in handling information. Alternative systems often are diversified, compounding farmers' needs for knowledge. Monitoring crops or livestock requires a sense of what to look for, which comes from both experience and insight. The information needed to manage a system may be too great for every farmer to absorb and use effectively. Although outside sources of help, such as magazines, crop advisers, and extension, are available to all farmers, some are more adept at obtaining such help and using it effectively.

Another reason to expect that alternative systems will require superior management is that generic or area-wide advice must be adapted to the farm's specific conditions (Janke and McNamara, 1988). Such tailoring requires the farmer to interpret and integrate different kinds of information, and it may call for some intelligent trial-and-error (Walters et al., 1990).

A related argument is that the components of the production process interact more strongly in alternative systems than in conventional systems. If so, maximizing the positive interactions would require sophisticated understanding. For example: "In conventional 'higher-input' farming, high yields can be obtained without appreciable attention to interactions. . . . However, as chemical inputs are lowered progressively, so the need for attention to the mechanism by which one input impacts upon another increases" (Edwards, 1987, p. 150).

Although these arguments seem reasonable, we do not know whether, in practice, alternative systems really require a better understanding of how the components of the agroecosystem interact. Often, in papers that make this argument (e.g., Francis and King, 1988; Pimentel et al., 1989), the examples offered are separable techniques that farmers can select individually, without having to integrate them into any kind of "system." Also, even though the structure needed to get the desired interactions may be complex, "many of the effects are common to well-structured systems. . . . [Therefore] total information . . . is not necessary to their implementation" (Harwood, 1985, p. 74).

MANAGEMENT ABILITY OF FARMERS WHO NOW USE ALTERNATIVE SYSTEMS

Given that these arguments are inconclusive, what do we know about the management abilities of farmers who actually use alternative systems? Unfortunately, very little. The limited study of this subject so far has been mostly anecdotal. Also, it often concerns organic farmers and others who have greatly reduced their chemical use, and may not apply to those who adopt less drastic alternatives. Case studies often comment that alternative farmers are excellent managers. But no doubt many other alternative farmers just muddle through; they are less likely to be selected for case studies. Nor are the excellent managers among conventional farmers likely to be subjects of case studies, as what they are doing is not especially unusual. Finally, farmers who now use alternative systems may be superior managers out of necessity, being a minority who must fend for themselves; they get little help from established sources of advice.

The idea that low input does not mean low-caliber management will appeal to farmers who want the satisfaction of making decisions and bearing the responsibility for what happens on their farms. Also, farmers are notoriously sensitive to what people say about their operations. The publicity given to some organic farmers' good management skills has helped counteract the unfortunate stereotype that farmers who choose not to use modern chemicals are backward and unsophisticated. Favorable perceptions played a similar role in the early days of the experiment station system. At that time, an expected side benefit of developing farming methods that required more skill and scientific understanding was that it would raise the image of farmers as professionals (Danbom, 1986a).

On the other hand, emphasizing management sophistication might backfire by causing farmers to worry whether they have the required skills. Some farmers would be more confident about switching to alternative methods that had been worked out to require less judgment or individual adaptation. Recognizing this, one organic farming advocate and practitioner has proposed breaking down alternative systems into separate elements that most farmers could learn easily (Coleman, 1985). Whether this is possible without seriously compromising the essential character of an alternative system is controversial: "Ecological agriculture has the potential of being presented to farmers . . . as a user-friendly package that does not have to be understood. . . . But difficult as it may be, ecologists will have to resist the temp-

93

tation to prepare simplified user manuals without serious explanations of what is going on" (Ehrenfeld, 1987, p. 185).

Those who are developing alternative methods need not choose between these positions. On one hand, the best results no doubt will be achieved by those farmers who are highly sophisticated in managing information actively. On the other hand, why not also try to develop "farmer-proof" methods, that is, methods with a good chance of success that many farmers would be willing to try, whatever their management ability? If research on alternative systems pays more attention to farmers' use of information, eventually we will know which strategy is more fruitful.

Information Requirements and Management
Ability in Conventional Systems

The alternative agriculture literature often says that conventional practices, such as heavy use of agricultural chemicals, override environmental variations. If so, site-specific conditions are less important and there is less need for monitoring and other information-related tasks.

This argument may be partly correct, but it can lead us to understate the information requirements of conventional systems. All farmers must continually gather and process information if their operations are to be profitable. Chemical use seldom means fixed applications that the farmer repeats mechanically every year, as its detractors sometimes claim. Even with conventional pest control and fertilization, information-intensive techniques allow farmers to use chemicals more efficiently. True, farmers sometimes apply chemicals at the highest rate that might be needed or even higher. But the margin between gross revenue and the cost of production often is so narrow that failure to select the most efficient rate could be economically ruinous.

Techniques to reduce or eliminate chemical applications in alternative systems sometimes are the same as those that farmers use to apply chemicals more effectively in conventional systems. For example, farmers can use field observations and an expert system to decide which insecticide to apply and exactly when to apply it to get the best control; such techniques are not used only to decide whether insecticides can be omitted. Similarly, a soil test

94

is done the same way whatever its purpose: to learn the economically optimal fertilizer rate in a full-chemical system or to learn how much to reduce the rate after using an alternative nutrient source, such as a leguminous green manure (Magdoff, 1991).

Even in conventional systems, the farmer must collect many kinds of information throughout each season. At the beginning of the season, choosing the best planting date and tillage method requires close attention to weather. During the season, efficient timing of irrigation requires frequent monitoring of soil moisture and the condition of the crop. For a crop that is dried artificially, the choice of the best harvesting date depends on balancing the cost of fuel against field losses, which in turn requires day-to-day monitoring of the weather and the crop's moisture content.

Also, good marketing imposes high information requirements in conventional systems. (The marketing challenges of alternative systems are sometimes given as a reason that they require superior management ability, but this applies mainly to systems with differentiated products sold through alternative channels, most notably certified organic.) For example, a cash grain farmer must keep track of commodity futures, domestic and overseas supply and demand forecasts, and availability of transportation. Getting the best price is especially important when the farmer produces only a few commodities; there is little chance to offset poor marketing of one by a good price for another. In contrast, the diversified product mix of many alternative systems provides some buffering against unpredictable price fluctuations, although it also requires the farmer to follow the markets for more commodities.

Implications for Research

INFORMATION HANDLING AS AN EXPERIMENTAL VARIABLE

Prevailing research approaches are not well suited for agricultural systems where the farmer's ability to handle information is critical. This shortcoming can be especially serious for alternative systems, which are thought to require more flexibility than conventional systems in adjusting to weather, weeds, insect pests, and other factors.

If researchers deal with farmers' management ability at all, they usually treat it separately from the system's biological and agronomic aspects. Agricultural scientists who are developing and testing a new method might look ahead at farmers' use of it—for example, in simplifying it for wide adoption. Often, however, they consider farmers' use of the method only after the research is completed, during the demonstration and extension phases. In a typical experiment, the management methods are specified in advance. But in actual production, the farmer might, for example, decide to cultivate an extra time if weed problems are unusually severe. If the farmer's ability to make such decisions is crucial to the system's success, research that takes no account of it will not be realistic.

STUDYING FLEXIBLY MANAGED SYSTEMS

The problem of incorporating management into research is illustrated by studies of organic farming. Besides avoiding conventional pesticides and fertilizers, organic systems typically differ from prevailing practices in other ways: more diversified crop rotations; greater reliance on livestock manure; and more frequent use of legume cover crops or green manures. Practical limitations rule out a multifactorial experiment that studies all these differences individually. Therefore, we should compare the performance of organic and conventional systems in their entirety rather than extrapolating from the systems' separate components (Shennan et al., 1991). The farmer is part of the system, especially an alternative system such as organic farming, where the farmer's adjustments to varying conditions are considered critical.

How should such a whole-system study be done? The results will be a more valid measure of the system's performance if the experiment responds to conditions during the growing season. However, the usual statistical methods work best with fixed treatments. In a multiyear study, varying the management from year to year may cause problems, such as confounding weather-related and treatment-related sources of variation.

A related question is how the researchers should decide on midseason adjustments that require judgment. This is different from research on an objective decision-making procedure, such as an expert system that recommends whether the farmer should apply an insecticide. There, too, the application rate is not known beforehand, but the "treatment," which is the method for choosing it, is fixed. However, organic and many other alternative systems cannot simply be managed by the book: the way in which the adjustments are decided, and by whom, are additional sources of experimental variation.

MAKING FARMERS PART OF THE RESEARCH

Researchers must deal with farmers' varying ability to handle information that is considered critical. At least this ability should be included as a confounding variable, similar to weather or soil type. Standard experimental procedures average over the latter variables, but not over management ability. Giving it its appropriate place presumably requires more research on working farms: farmers' judgment is not a factor that researchers should try to simulate. By implication, working farmers should be part of research earlier than is now customary. We should not deal with them only after the research stage is finished and the techniques are ready to be moved off the experiment station. Since prevailing disciplinary boundaries largely separate the human aspects of agriculture from its technical and biological sides, such research would involve several disciplines.

Also, it is fruitful to make a distinction that sometimes is overlooked: a study that evaluates how well a system works differs from one that tries to explain why the system works as it does. Both are legitimate, but they often require different experimental designs and sites (Lockeretz, 1985). To learn how well a system works, we should study that system, not an artist's conception. In particular, if adjustments during the growing season are part of the system, they should be made part of the research too. In a whole-farm study done on a working farm, the researchers have no choice but to include such adjustments. On the other hand, the research may be intended to learn the detailed relationships among components of the system. Although keeping certain practices fixed may be unrealistic, in this case it might be necessary so that "irrelevant" variation is controlled.

Experiment stations often will be more suitable for explanatory studies, whereas working farms seem especially appropriate for evaluations. Also, evaluations should be done on more than one farm, to get variation both in

97

growing conditions and in farmers' management ability and decision-making styles. If the prevalent view of alternative agriculture is correct, judgment and insight are sufficiently important that even experienced farmers might make different decisions, given identical conditions.

Information and Management Ability:
Essential in All Production Systems

Alternative agriculture provided the stimulus for this discussion of information-handling and management ability, but the issues we have raised apply more broadly. Depending on the production system, the farmer will need different kinds of information, but the success of any system demands that the farmer continually gather and think about new information. The difference between conventional and alternative systems, if any, is one of degree.

True, information and decision making are especially likely to come up when people talk about alternative agriculture. But we should not conclude too much from such talk: the role of information and decision making in conventional systems probably has been understated. When some people promote alternative agriculture, there is an undertone of disdain—sometimes more than an undertone—because farming supposedly is so easy when chemicals are used freely.

This claim is a great oversimplification. The main difference may be that for alternative systems, information is explicitly recognized as a way of replacing material inputs. In contrast, in a conventional system it simply may not rate special comment.

Failure to address the role of information explicitly is a serious shortcoming in how we analyze agricultural systems. No system, whatever material inputs it uses, can achieve its potential if intangible inputs like information are not given as much careful thought as material inputs. The talk about the information requirements of alternative agriculture has reminded us of that. Its most significant benefit could be that it makes us pay more attention to an input that is important in every agricultural system, but that we sometimes shortchange in research. If so, the concept "information-intensive agriculture" will have value far beyond what it contributes to alternative systems in particular.

8

. . .

On-Farm Research

I take it for granted you agree with me that cooperative work between the station and the farmer is desirable. . . . But to what type of investigations should it be confined, and how should cooperative work be carried on? We shall probably agree that certain kinds of investigation work can be conducted only in the laboratory or on the station grounds and under the eye of the station worker. The results thus obtained frequently require verification and demonstration by trials made on a larger scale and under less ideal conditions. In most cases they also, from the standpoint of the farmer, need practical illustration. . . . Experiments in which the farmer cooperates with the scientist . . . [have] considerable advertising value. The results, be they negative or positive, are the more readily diffused in that immediate locality.
—Craig, 1902, p. 102

Doing more research on working farms has been advocated by many groups, especially those interested in alternative agriculture. For example, the pioneering national grant program on sustainable agriculture gave "high priority . . . to on-farm research and demonstration projects that provide for scientific documentation of low-input sustainable farming practices and systems" (Sustainable Agriculture Operations Committee, 1991, p. 4). Similarly, the National Research Council (1989b, p. 23) remarked that "on-farm research will have to be increased and directed toward systems that achieve the multiple goals of profitability, continued productivity, and environmental safety." On-farm research has long been a cornerstone of farming systems research in the developing world (Hildebrand and Poey, 1985, chap. 1). Interest in it in this country has arisen independently, especially among farmer groups.

Two aspects of on-farm research are often intertwined. One is the influence it gives to farmers in deciding what topics should be studied and how. The second is the advantages that farmers' fields offer as research sites (Anderson, 1992). To some people, farmer-controlled and on-farm research are so closely linked that the distinction makes no sense. We prefer to discuss them separately, however.

The influence it gives farmers is an important reason that some family-farm advocacy groups have demanded more on-farm research. They are dissatisfied with the direction of contemporary agricultural research, which they see as favoring large farms and agricultural input suppliers to the detriment of smaller farms and the environment (Burkhardt, 1991). We defer this issue to the next chapter. Here we deal with the choice of site as a matter of research strategy. The question is: How can research benefit by using farmers' fields?

To answer that question, we first clarify the differences in the forms of on-farm research. We then explain why such research is associated with alternative agriculture and why it also is valuable for a broader range of topics. We discuss additional advantages that sometimes favor putting a project on a working farm even if it is suitable for an experiment station; these include benefits for the researchers, for farmers, and for demonstration projects. Finally, we analyze problems that may arise because of the need for collaboration among different groups.

Different Styles of On-Farm Research

On-farm research comes in many forms. Among the important differences are:

- How control is divided between farmers and researchers;
- Whether the primary beneficiaries are farmers or researchers;
- Whether the experimental site is isolated from or coupled to the rest of the farm;

• Whether the research imposes its own treatments or studies the production methods already used on the farm.

Although interest in it is growing, on-farm research is not a new idea. For example, crop variety trials have long been done on farmers' fields. The older style of on-farm research usually examines questions similar to those studied at experiment stations, using similar procedures. That the research is done on a working farm is not fundamental: the farmer's field simply is a piece of land that happens to offer a soil type or some other condition not available on the experiment station. The researchers may take pains to isolate the experimental area from other activities on the farm. Similarly, the farmer usually has only a small role.

NEWER KINDS OF ON-FARM RESEARCH

In contrast to variety trials, the newer kinds of on-farm research deliberately take advantage of the fact that the research site is part of a working farm. In the version called "farmer-managed research," the investigation is planned and carried out by a farmer. Its purpose is usually to solve a particular production problem or to evaluate an innovation under the conditions of that farm. The results are usually publicized through field days, newsletters, or farmers' networks. Professional researchers at most provide help with experimental design and data analysis. A prominent example is described by Thompson and Thompson (1990); similar programs are common in many other countries, both industrialized and developing (Elkana, 1991).

These approaches do not exhaust the possibilities. Projects in which farmers are heavily involved can still be like traditional research in other respects. First, professional researchers can retain considerable responsibility. Second, the results can be published through customary professional publications, such as journal articles and experiment station reports, although farmer-oriented channels might be used also.

This chapter deals with on-farm research in which researchers retain much of their traditional professional role because we are interested in how farmers' fields can complement or substitute for an experiment station. This is an issue for researchers, but not for farmers, who do not have a choice of sites.

Newer styles of on-farm research, even if initiated and controlled by researchers, are different from the older style that treats the farm like an extension of the experiment station. For example, the researchers might not intervene at all in how the experimental site is managed. Instead, they simply make measurements on a field managed as part of the farm. More often, researchers and farmers work collaboratively, using a design that compromises between the nonintervention approach just described and the older, more station-like type. The site receives experimental treatments, but the farmer does at least some field work, using full-scale equipment. Also, the treatments are chosen in consultation with the farmer. Instead of trying to insulate the site from its setting on a working farm, the experimental procedures are selected with consideration of the constraints that govern real farms. Allowing such constraints to come into play is a good reason to do research on working farms.

THE STATUS OF THE NEWER KINDS OF ON-FARM RESEARCH

Some researchers have pursued the newer styles of on-farm research enthusiastically. Others, however, have been reluctant to accept them (Anderson, 1992). They consider them less rigorous and credible, less able to determine causal relationships, and more subjective because they are assumed to be based on informal observations, not on systematic data collection.

These assumptions are unwarranted: on-farm research, like station research, can have any degree of objectivity, and is fully compatible with hypothesis-testing statistical methods. The notion that it always is "softer" than the "hard" science done on experiment stations is a mistake.

Even among researchers who acknowledge the full scope of on-farm research, questions remain about the validity of various methods. Accepted methods have emerged from farming systems research (e.g., Hildebrand and Poey, 1985), but this approach has been used primarily for subsistence farming in nonindustrialized countries. Researchers in the United States have tested and validated experimental designs appropriate for some kinds of on-farm trials (e.g., Rzewnicki et al., 1988; Shapiro et al., 1989). Such procedures usually are modifications of standard experimental designs, differing mainly in plot size and the number of treatments and replicates. Work is needed on accepted methods for comparing whole farms and entire production systems and for research in which farmers' management skills are part of the system being studied.

The Association between On-Farm Research
and Alternative Agriculture

On-farm research, which can give farmers more influence on research topics, is especially popular with many proponents of alternative agriculture. Without it, alternatively oriented farmer organizations might not have much influence on experiment stations.

Another reason for the association has to do with alternative agriculture's reduced use of chemicals. Reduced-chemical systems are presumed to be more sensitive to environmental variations because they depend more on natural processes for controlling pests and supplying nutrients. The small plots and artificial conditions of an experiment station may not accurately capture these systems' coupling to the environment. In contrast, chemicals are considered to override some environmental variations; to take an extreme example, if we wipe out the soil's microflora by fumigation, we need not concern ourselves much with variations in pathogen abundance.

The role of the farmer, which is presumed to be more critical in alternative agriculture, can best be studied on working farms. Alternative systems are commonly described as requiring more management sophistication and greater use of information than conventional farming systems. That is why it has been said that "fiddling with these systems on each farm by individual farmers is the primary way to get them to work" (Janke and McNamara, 1988, p. 136), and that "successful conversion usually requires that farmers become researchers, and their farms become experimental farms" (MacRae et al., 1990, p. 161). Similarly, for researchers seeking to understand such systems, information on farmer management "can only be obtained by on-farm research" (Edwards et al., n.d., p. 4).

A related point is that alternative systems often are assumed to involve the farm as a whole, so that they cannot be developed simply by changing individual enterprises or production techniques. Instead, they should be investigated on working farms to allow the researchers to observe interactions among their components. Moreover, some alternative methods were first used by farmers; their farms are the place to start investigations, so that researchers can take advantage of these innovators' experience and creativity (Krome, 1988b).

An additional link with on-farm research arises because supporters of alternative agriculture emphasize what happens beyond farmers' fields, including farmers' social relations and the well-being of farming communities. Public research institutions traditionally have not accepted responsibility for anticipating the social consequences of new production systems: "These concerns . . . often have been viewed as peripheral or diversionary" (Ruttan, 1991, p. 121). If this neglect is to be corrected, the broader effects of production systems need to be examined directly. Working with farmers on real farms is critical for assessing the social effects of farm practices, since complex social relations in rural communities cannot easily be simulated.

Despite the validity of these arguments, they do not mean that experiment stations and other traditional research sites, such as greenhouses and growth chambers, have no role in research on alternative systems. The realism of an on-farm site means sacrificing control over important experimental variables. A balanced research program will include studies done at various points along the "controlled" to "realistic" spectrum. Whole-farm and "systems" research must be supported by good studies of the system's components; for that, an experiment station often is the best site.

Also, some research necessary for developing alternative systems is not connected with a specific production system (chap. 4). Alternative agriculture can benefit greatly from research in plant physiology, soil chemistry, genetics, insect physiology, and many other fields. In promoting research under realistic farm conditions, one should not forget the value of research that does not take place on a farm of any kind, commercial or experimental.

Finally, on-farm research is constrained by the area's prevailing agricultural structure: it cannot study a system requiring a type of farm that does not exist, such as a mixed crop-livestock farm in an area dominated by cash grains. Ironically, the "alternatives" to be studied on-farm cannot differ much from what some farmers are already doing. On-farm research has been conservative, dealing mainly with small changes in practices. Typically, it accepts the prevailing farm structure as given, even though structural factors like farm size and specialization may be at the root of the environmental and socioeconomic problems it hopes to solve.

The advantages of using farmers' fields also apply to research with a conventional orientation. For example, even heavy use of chemicals does not offset all environmental variations, and results may change when the experiment is transferred from the partially unrealistic conditions of an experiment station to a working farm. Second, farmers' management is significant in many systems, alternative and otherwise, and farmers have been innovators in many ways. Finally, the off-farm consequences of any production system matter, even if the system was not designed with such effects in mind. If anything, these effects deserve even more attention, a lesson we should have learned from many previous instances of new production systems that caused unforeseen socioeconomic or environmental harm.

For these reasons, we discuss the advantages of using farmers' fields independently of whether the research is "alternative." Working farms may be especially suitable for alternatively oriented research, but the choice of site still depends on the particular question the research addresses, not on its general orientation (Lockeretz, 1987).

Research Appropriate for Working Farms

RESEARCH THAT MUST BE DONE ON WORKING FARMS

On-farm research is essential under three circumstances:

- The necessary conditions are not available on an experiment station.
- The way farmers manage the system is part of the research question.
- The study requires a whole farm.

Site-related conditions. Farmers' fields offer a wide range of soil types and other conditions. Capitalizing on this variation is especially valuable in a large, heterogeneous state that does not have branch stations in all its important agricultural regions. Also, researchers may need to study a natural phenomenon that should not be reproduced on the station, such as a pest problem or disease.

Sometimes the research concerns a practice or cropping system that would take too long to establish on the station. That is why studies of organic farming's long-term effects on soil often have been done on working farms (e.g., Lengnick and King, 1986; Reganold et al., 1987; Shennan et al., 1991). Also, the topic might require an experimental design that is incompatible with experiment station constraints. For example, a study of the life cycle and movement of an insect pest should be done on the scale over which the insect moves, which may be too big for an experiment station. Similarly, a multicrop, multitreatment rotation study may need more land than is available on the station. A large area may be needed to study runoff and erosion.

Management. In the previous examples, the site itself was the reason to use farmers' fields. Another group of questions must be studied on farms because management is part of the study (Thornley, 1990). If farmers are using an unfamiliar innovation successfully, an on-farm site lets researchers learn more about it before they study it on the station. The problems that can arise when this step is bypassed are illustrated by a "low-input, high-management" system set up for comparison with two other systems at a research center. The investigators could have avoided its low yields and heavy weed pressure by working with local farmers who were already using it successfully (Exner, 1990).

Researchers may want to compare results from farmers' fields with those obtained in researcher-managed plots on the station, or to learn how results vary among farmers with different management styles, as we suggested in the previous chapter. Finally, researchers must go to farmers to study farm decision making and its subjective underpinnings.

Whole-farm studies. The third kind of on-farm study involves whole-farm dynamics. If researchers are interested in interactions among the farm's enterprises, they usually must study working farms. Different enterprises can be set up simultaneously on an experiment station, and economic simulations can be constructed from data obtained from multiple sources (Dobbs et al., 1990). However, whole-farm biological interactions cannot be studied off the farm. At best, an experiment station can look at these interactions among a few enterprises.

As a compromise, a portion of an experiment station may be managed as a complete farm or an existing farm may be dedicated to experimentation. The merit of this option is arguable. On the positive side, it allows researchers to study whole-farm interactions in a controlled setting. On the negative side, this degree of control is what makes a model farm unlike a real farm, creating problems in applying the results off-station. Before making such a commitment, station administrators need a convincing reason for duplicating conditions already available on local farms, instead of doing research that can be done only on the station.

A nonreason for on-farm research. A common but invalid reason to use farmers' fields is pressure from funding sources. When program guidelines mandate the use of working farms, the quality of the research may suffer even though the intention is good. The choice of site is a decision that calls for case-by-case judgment, not rigid prescriptions.

ON-FARM RESEARCH AS PART OF A LARGER RESEARCH EFFORT

On-farm research can complement research conducted on-station; the choice is not limited to one or the other (Anderson and Lockeretz, 1991). The station often is better for studying how a promising innovation works, as opposed to evaluating how well it performs. Also, station research is suitable for trying riskier alternatives. Farmers are wary of experiments that may make a field look bad or that will leave them with a lingering weed or disease problem. After investigations on the station, techniques that seem suitable can be tested on working farms before researchers recommend them to farmers.

Another way that station and on-farm research can support each other is for both to contribute data to whole-farm simulations. However, there may be systematic differences between farm and station data. For example, grain

107

yields from large, machine-harvested plots generally will be lower than yields from small, hand-harvested station-type plots, because machine harvesting misses lodged stalks. Also, either farmers or experiment station personnel may be better at using a particular production method.

The complementary roles of on-farm and station research are related to the complementary roles of whole-farm and component-level studies, but the two issues are different. On-farm research often is assumed to be whole-farm research, but it can focus instead on specific techniques or enterprises, as station research typically does. Conversely, working on a station does not preclude a whole-farm focus; as we have noted, data from station plots can contribute to whole-farm models.

Even on-farm research on a single field can have a whole-farm aspect. If farmers do the field work, using their own equipment, the site is managed as part of the farm and is subject to whole-farm constraints. The farmer needs to make a profit from a farm, and work on the research plots must fit in with the other jobs the farmer must do.

Some investigators combine the two scales of on-farm work to good advantage. They collect data intensively from specific sites, but also from the farm as a whole. Typically, microsite data concern biological and physical variables, such as soil flora and fauna, whereas the whole-farm analysis is usually economic.

Additional Benefits of On-Farm Research

BENEFITS FOR RESEARCHERS

A project can be enhanced by being done on a farm because the farmer can help in designing it. The farmer, who does not necessarily share the researchers' assumptions, can suggest interesting questions that researchers otherwise might have overlooked. The farmer's advice in planning can help make the research more practical, efficient, and useful to local farmers (Granatstein, 1988; Krome, 1988a).

Beyond the planning stage, researchers can benefit from what farmers contribute during the course of the project. A farmer may notice peculiarities that the researchers miss. The farmer, who knows the farm well, may suggest why a practice works well or fails in a particular field. Because the farmer is present on evenings and weekends, on-farm plots can receive more continuous attention than experiment station plots. Also, farmers and researchers often observe the same phenomenon differently. The difference in perspective can be frustrating, but it also can be valuable in interpreting the observation. On-farm research is not the only setting where this happens, but it has long been recognized as effective:

> This work, carried out in their own locality, under local climatic, geographic, and agricultural conditions in their own fields . . . is actual experimentation with the farmer taken into partnership. . . . Many farmers will sit through an institute meeting, listening intently, but will ask no questions and give no experiences. Somehow it seems as though a body of people brought together in this way gives the average farmer a species of lockjaw. Yet these same men, interviewed in their own fields by someone who fits in with their life, immediately reacquire the power of speech and give out information freely, often supplementing the knowledge acquired by the entomologist in his laboratory. If these field stations accomplished nothing more than this they would repay over and over again the funds annually appropriated for the work. (Webster, 1914, pp. 88–89)

Finally, on-farm research offers advantages even if the farmer is not highly involved in the work. A researcher who did not grow up on a farm can gain useful experience and appreciation of farmers' constraints by working on-farm. It also can improve the public image of researchers and extension workers and make the research more understandable to decision makers (Hildebrand and Poey, 1985, p. 5).

COSTS AND LOGISTICAL CONSIDERATIONS

An on-farm site sometimes reduces researchers' costs. However, such lower costs are not a general rule, for two reasons. First, the cost comparison depends on who does what. Besides the research site, farmers sometimes provide equipment and labor worth more than the compensation they receive, if any. On the other hand, on-farm work can be more expensive when researchers supply all the equipment, labor, and inputs. Extra time to coordinate a project also is expensive.

The second difficulty in comparing costs is that the two sites may not provide comparable information. Intensive monitoring that requires special technical skills or equipment probably should be done on-station. On the other hand, if the farmer can do the monitoring, a farm is advantageous because the farmer is continuously on the site and does not have to commute a long distance to the study area.

On-farm research sidesteps several logistical problems of station work, such as coordinating the use of equipment or competing with other researchers for land. However, it introduces new logistical and communication problems. In a complex, multidisciplinary experiment, many scientists and technicians must coordinate their activities, not just among themselves but also with the farmer and perhaps a farm manager and a crew of workers.

BENEFITS FOR FARMERS

Some groups advocate on-farm research mainly because of its intangible personal or community benefits. When farmers participate in research, they must examine their own decisions, perhaps becoming aware of options they had not seen before. They learn something about research methods and may continue the experiments independently.

This point was recognized early, as in this assessment of farmer-conducted fertilizer studies organized by the scientist who later became the first head of USDA's Office of Experiment Stations:

> Such experiments . . . [are valuable] for the information they bring, yet, over and above this they have a still higher usefulness in the stimulus they give to closer study, more accurate observation and more rational application of the principles of science. The apparent object in introducing them was to work upon farmers' soils. Underneath this lay in my own thought, a deeper purpose, to work upon their owners'

minds. And in this regard, at least, the outcome has been most gratifying. (Atwater, 1878, pp. 366–67)

One cooperator even credited the experiments with keeping him in farming, commenting that "[I was] supplementing the labor of my hands with the labor of my brains, and I feel the benefit in my purse, in my home, and in my mind" (Atwater, 1882, p. 363).

A farmer who participates in research is more likely to change to or continue using an improved practice after the project is over. Working on research with other farmers can build community ties and lead to more cooperation with neighbors, such as sharing information, labor, and equipment (Nopar, 1990).

BENEFITS FOR DEMONSTRATION PROJECTS AND FARMER EDUCATION

As we have noted, some people have biased impressions of the scope and quality of on-farm research. Their attitude may result from a failure to distinguish on-farm research from demonstrations, a related activity with a different purpose (Lockeretz and Anderson, 1990).

Demonstrations must be visually convincing, but they need not include data collection or analysis. Often they are done at a single site, and they may be short-term and nonreplicated. These factors make them easier and cheaper than research projects. They can stimulate farmers' thinking about different systems and convey simple recommendations. But unlike research, demonstrations do not contribute to a cumulative body of knowledge.

In deciding whether demonstration or research is appropriate, project planners need to keep in mind the audience for whom the work is done and the respective purposes of the two activities: research seeks to answer a question, whereas demonstrations seek to educate and persuade. Unfortunately, the lower status often imputed to demonstrations may tilt the choice toward research, even when a demonstration would be more valuable. Also, a demonstration may be labeled research because of the "halo" surrounding the term (p. 40).

Sometimes the same activity can serve both purposes. Research that also is used for demonstration may be more credible and accessible to farmers when it is done on farmers' fields instead of an experiment station. Some farmers are dubious about the small plots typical of station research. Also, if a farmer manages the site, the demonstration is closer to what other farmers can expect on their own farms.

Another advantage of working with farmers in combined research-demonstration projects is that it helps insure that the project meets farmers' needs. This is not the only justification for research, but it certainly should be a criterion for any research that also is the subject of a demonstration. Cooperating farmers can advise researchers whether other farmers would be likely to use a method being considered for research-demonstration. They also can help disseminate research results. If the research involves a difficult production technique, the participating farmer can coach other farmers as they begin to use it.

Project Organization and Personnel

Success in on-farm research requires effective cooperation among people with diverse interests and backgrounds. Researchers must come to an agreement with the farmers, at least on use of their land. People from several disciplines may have to collaborate, especially when whole-farm phenomena are studied or the research is concerned with alternative agriculture. Also, many funding sources that support on-farm research encourage cooperation between research institutions and outside groups, such as grass-roots organizations.

COLLABORATIONS AMONG DIFFERENT KINDS OF GROUPS
Groups involved in on-farm research have different perspectives, corresponding to their different strengths:

• Farmers can offer their experience, their opinions about what problems are important, and their judgment about whether an alternative is practical. They also may supply labor and equipment, along with land.

• University personnel have training in specialized disciplines and research methods. This enables them to judge the reliability of different kinds of studies, interpret their applicability to different conditions, and integrate them with other research results.

• Extension personnel have experience working with farmers and have contacts in every county.

• Private grass-roots organizations and state programs in alternative agriculture, unlike extension, need not serve a specific quota of farmers. They can thus spend more time with individual farmer-cooperators. Also, they tend to emphasize topics that are not already getting sufficient attention at land-grant universities. This can help insure that the total research effort in a state meets the full range of public needs.

Partly because of their different views of research and their different constituencies, these groups may have trouble working together. Team members must respect each other and be able to communicate well. Each group must see how its purposes will be served and must be open to sharing what once was exclusive control. New private, nonprofit organizations and state programs may arouse hostility or suspicion from the extension service, which already has good working relations with farmers (Anderson and Lockeretz, 1991).

In the case of an unconventional practice, it may not be possible to get the needed cooperation from extension agents because they are not familiar with the practice. Alternative agricultural systems have largely been outside the mainstream, and their practitioners have not relied much on extension personnel in the past (Baker and Smith, 1987). On the other hand, involving extension agents lets a community know what is being done and helps overcome the resentment the agents otherwise might feel against someone they perceive as encroaching on their territory.

Another source of conflict is different ideas about the purposes of a project. Sometimes these differences become apparent only after the work is in progress. For example, a private organization may do research because it wants to foster social change, whereas university researchers may believe that promoting a particular viewpoint jeopardizes their professional credibility.

Collaborative on-farm research demands extra time and disciplinary flexibility. Junior researchers may avoid such a project because they feel obliged to publish frequently in disciplinary journals. Also, they are more vulnerable to repercussions from working with nonprofessional groups, although this reaction is becoming less of a problem as administrators realize that cooperation with such groups can foster valuable political support for the university.

COLLABORATION AMONG DISCIPLINES

The system-level and whole-farm scope of much on-farm research implies that it will be multidisciplinary. Moreover, both multidisciplinary research and on-farm research are especially associated with alternative agriculture. This link makes on-farm research even more likely to be multidisciplinary.

A special issue in multidisciplinary on-farm research is the need to compromise on a site (Shennan et al., 1991). Farmers' fields offer a greater variety of sites than an experiment station. Researchers from different disciplines often seek different sites to do their parts of a project properly: an entomologist, for example, might want a site where an insect outbreak is especially severe, but the factors that caused the outbreak might make the site unrepresentative for other team members' work.

WORKING WITH FARMERS

Farmers are not necessarily involved in on-farm studies beyond giving permission to use their land. Their fields may be used only to take advantage of particular conditions or management history. However, farmers' participation is vital when what they do—not the land they happen to own—is the reason for working on their farms.

Farmers' contributions can improve the planning and execution of on-farm research. Before researchers make firm plans, it is valuable to get farmers' advice on which practices deserve attention and how the study plots can be kept compatible with the farms' constraints. Also, farmers are more effective educators about project results if they have been involved in designing the work. Finally, farmers' participation enhances the personal and community benefits.

On the other hand, the experiment may require more sophisticated monitoring than the farmer can provide dependably. A drawback of farmers' heavy involvment is researchers' lack of control over procedures. A familiar story in on-farm research concerns something that goes wrong on the experimental site while researchers are absent. The reason may be poor communication or a need that had greater priority for the farmer than maintaining the experimental area the way the researchers wanted.

Mistakes happen at experiment stations too, but they are less likely to cause a total loss. Although such risks cannot be eliminated from on-farm research, several precautions can help prevent it: farmers' and researchers' respective responsibilities must be clear; researchers must adapt the experiment to what they judge each farmer can do; and farmers should have some sense of "owning" the research.

Appropriate Expectations for On-Farm Research

Research done on working farms, especially with high farmer participation, is increasingly being recognized for its contribution to improving our agricultural system. On-farm research that takes account of farmers' management and that includes off-farm effects can help researchers design systems that are both more efficient and environmentally benign.

What to expect from on-farm research should not be overblown, however. For research that tries to explain why a new system works the way it does, an experiment station often is a better site. Also, on-farm research will not necessarily give farmers greater political power, nor insure that the benefits of agricultural research go primarily to family farmers. These are appropriate goals for advocacy groups, and they deserve the attention of researchers too. But they will not be achieved simply by putting more research on working farms. The decision to do a project on working farms and to involve the farmers should be based on serving the goals of the research and enhancing its quality.

9

. . .

Farmers' Influence on Research Priorities

Many of our station workers see only the immediate duty of the station to the local farmer of to-day. They forget that the station has a duty to all phases of agriculture in a broad sense, in order that its labors may lead to much more permanent and wide-spread benefit.—Committee on Experiment Station Organization and Policy, 1907, p. 75

Many alternative agriculture proponents have called for farmers to have more influence in choosing research topics. Sometimes they go further, advocating a greater role for farmers in designing, carrying out, and evaluating specific projects: "[Alternative agriculture research] should involve farm families in all possible stages. . . . Farmers should be included in an advisory capacity to the overall research program" (Krome, 1988b). Similarly, when USDA established a national research program on low-input agriculture, a "guiding principle" was that "meaningful participation of operating farmers . . . is essential to the success of this approach" (Madden and O'Connell, 1989, p. 9); moreover, the program put this principle into practice (Schaller, 1991).

Advocacy of a greater role for farmers is not limited to alternative agriculture. One proposal is that "as a start, we should require that every agricultural research project be reviewed by a farmer prior to funding approval" (Thornley, 1990, p. 176).

Many researchers recognize that farmers have much to contribute to research decisions, a turnaround in an attitude that once was more prevalent (Gardner, 1990). Traditionally, farmers have been regarded as the passive— initially, even reluctant—recipients of research results (Danbom, 1986b). They might influence priorities in a general way, but rarely were they involved in planning or evaluating specific projects.

The current call to increase farmers' influence on research topics is a healthy challenge to the view that researchers have nothing to learn from them. A greater role for farmers will allow researchers to draw on farmers' experience and knowledge and to understand better their attitudes toward the systems being investigated.

Their participation, however, does not automatically mean that farmers should have an influence on each stage of every research project. If researchers take their cues primarily from farmers' expressed preferences, the result can be in conflict with the social goals of alternative agriculture. To help deal with this conflict, we will take up two questions.

First, whom is agricultural research supposed to serve? The answer includes farmers, of course. A more specific question then arises: What relationship between farmers and researchers will enable research to best meet its obligations to farmers, while also meeting its obligations to society as a whole? We do not believe that researchers should simply say to farmers, "Tell us what we should do." Rather, a balance must be sought between the complementary roles of the two groups. Such a balance would be preferable to either the "leave us alone and just let us do our job" elitism that some researchers exhibit or the equally one-sided "farmer-first" slogan offered as an alternative.

Discussions of farmers' influence over research priorities often tie it to on-farm research. Participating in research on their own farms makes farmers a kind of partner to the researchers. However, farmers' influence over research is a separate matter from the setting of the research. A working farm can make sense as a setting even when the farmer does not have a major role in making decisions. Conversely, farmers can influence research that is not done on their farms. For example, grant programs sometimes have advisory committees that include farmers (chap. 12).

To keep the distinction clear, the previous chapter was concerned only with on-farm research as a strategy for achieving research goals. Here we deal with farmers' involvement in choosing those goals, whether or not the site is their farms.

Farmers' Research Needs and the Goals of Alternative Agriculture

The idea that farmers should have a greater role in choosing research topics is often linked with alternative agriculture. Even so, we should not take for granted that this role will serve the purposes of alternative agriculture research. Limited experience to date suggests that it may not.

Alternative agriculture is supposed to consider the long-term consequences of production methods and give high priority to reducing adverse environmental and socioeconomic effects. Also, alternative systems are usually characterized as based on complex agroecological principles (Stinner and House, 1987).

These characteristics mean that in aggregate, although not necessarily in every project, alternative agriculture research should differ from conventionally oriented research in several ways (Anderson and Lockeretz, 1992):

• It should run long enough to evaluate how a system will affect future agricultural productivity; for some effects, such as changes in soil organic matter, this could require more years than measuring profitability, say.

• It should extend beyond the farm's boundaries to the local environment of which the farm is a part, and beyond that to the larger economic system (Lowrance et al., 1986; National Research Council, 1989b, pp. 22–23).

• It should strive to understand the interactions among the elements that make up a farm: its various enterprises and production activities, and its soils and biota (Edwards, 1987). The hope is that this understanding will lead to improved production systems that advantageously integrate a farm's components (Francis et al., 1986; Stinner and House, 1989).

Yet alternative agriculture research programs—many of which give farmers an important role in choosing research topics—typically do not deal with long-term or off-farm effects on environmental quality, resource depletion, or rural communities (Anderson and Lockeretz, 1992). No doubt

farmers are concerned about these issues. In what they actually study, however, farmer-oriented alternative agriculture research programs sometimes
ignore the same questions that alternative agriculture proponents criticize
the research establishment for ignoring.

Instead, alternative agriculture programs typically deal with tangible,
short-term problems that farmers hope to be able to solve soon, such as a better way to control some weed. The short-term approach works against
studies of systems that require a major change in farm structure, such as a
more diversified enterprise mix. In other words, it reinforces the conservative influence of working on farmers' fields.

Another limitation of some alternative agriculture studies is that they simply compare the performance of various techniques or materials. Rather than
developing a better way to do the job, the goal is to enable a farmer to select
the best way from those that are already available.

Such trials are similar to what farmers demanded when the experiment
station system was established under the Hatch Act in 1887. Focusing on
those demands seriously cut into the system's scientific contribution (Ferleger, 1990). To free it to fulfill its scientific potential, later legislation (the
Adams Act of 1906) specified that the system's mission was to do "original"
research. Its federal overseers interpreted this as excluding simple comparisons of various products (Committee on Experiment Station Organization
and Policy, 1907; Rosenberg, 1964).

There is a way to avoid this tradeoff. Farmers can do more of the comparisons themselves, as we discussed in the preceding chapter. Alternative agriculture research can then be less directly tied to farmers' immediate needs
without leaving those needs unmet.

The Beneficiaries of Publicly Funded Agricultural Research

Some discussions of farmers' influence on agricultural research take as
given that the purpose of research is to serve farmers. If so, farmers clearly
should have the main voice in deciding research priorities; the only question
is how to incorporate their voice into the research process. However, if
farmers' research needs can conflict with broader alternative agriculture

goals, as discussed in the previous section, the fundamental question is: Whom is public research supposed to serve? This apparently simple question has complexities that are not always acknowledged.

The presumption that the public research system should serve mainly farmers has strong historical support. Publicly funded agricultural research at the land-grant schools—the "people's colleges"—was started with a clear commitment to serving farmers.

The meaning of "serving farmers" was not clear, however. It did not simply mean doing things that would help farmers. The people who established the system "spoke of doing something for the farmer, but their main goal was a productive agriculture which would strengthen the nation" (Danbom, 1986b, p. 110). It is interesting that this distinction was clear from the start even though the country was overwhelmingly agricultural at the time.

"Doing something for the farmer" definitely did not mean doing research that farmers said they wanted. If this had been the case, the research system probably would not exist. Nor did it mean involving farmers in the details of research. Very early, an understanding was reached in which researchers kept control over the research process but won farmers' support by emphasizing highly applied work of immediate utility to farmers (Schweikhardt and Bonnen, 1986).

Despite their initial antagonism, farmers now accept the idea of USDA and land-grant university research on their behalf. True, they retain some ambivalence, especially toward federal research that increases aggregate production and thereby lowers commodity prices (Buttel and Busch, 1988; Gillespie and Buttel, 1989). Still, farmers are largely the system's clients— in the view of some, the primary clients.

Nevertheless, controversy continues over the meaning of "serving farmers." Serving certain kinds can mean harming others, as in favoring large farms at the expense of small and moderate-size farms. By its selective interpretation of "serving farmers," the research system has helped transform U.S. agriculture to fit its notions of what is desirable (Kirkendall,

1986). For example, the research establishment has been criticized for continuing to emphasize increased production after that no longer was the best way to serve either farmers or the public (Burkhardt, 1991).

WHO PAYS AND WHO BENEFITS?

The money for publicly funded agricultural research comes from the public as a whole, and everyone is affected by both its intended and unintended results. To say that farmers are the main beneficiaries means that everyone else is being asked to spend a considerable amount per beneficiary. Historical precedent is powerful, but can farmers expect their traditional client status to continue indefinitely despite their declining political power? Is there an entitlement to a large research establishment dedicated to meeting farmers' needs, and if so, how well is it likely to withstand federal and state budget pressures?

The conventional answer to the challenge implicit in these questions is still, as in the early days of publicly funded research, that the nation as a whole is the real beneficiary. No doubt it is, but one can't have it both ways: if one claims that its public benefits justify asking the public to pay for agricultural research, one cannot also argue that farmers should have the main voice in shaping that research because they are the ones it is supposed to serve. One might argue that what is good for farmers is good for the nation, but agrarian fundamentalism is not likely to persuade a predominantly urban society.

The question "For whom?" has an ironic twist when it is applied to alternative agriculture. Much of the current interest in the subject and much of the willingness to spend money on it were stimulated by groups outside farming. Even before the term "alternative agriculture" became widely used, its ideas were advocated by environmentalists, people concerned about toxic residues in food, and rural activists interested in broad questions of social and economic equity. These questions were the "new agenda" issues of the 1960s and 1970s (Paarlberg, 1980, chap. 5) that the "externalities/alternatives" coalition promoted so effectively against strong opposition from some farmers' groups (Hadwiger, 1982, chap. 8). If research priorities really were chosen according to farmers' preferences, how much progress would alternative research have made by this time?

Of course, there are also many farmers who do favor alternative agriculture. But if we assert as a general principle that agricultural research is mainly supposed to serve farmers, that should mean *all* farmers. Farmers vary widely in the size and capitalization of their farms, their goals and values, and their ability to express themselves. Service to farmers should not be limited to those farmers who are most articulate in getting their wishes across to researchers, any more than it ever should have meant only those who could exert economic and political leverage through well-organized lobbying groups. We should not invoke the cause of "serving farmers" if in fact we mean only "the right kind of farmers" or only "farmers who want the same things we want."

OTHER GROUPS INTERESTED IN RESEARCH

One answer to the question "For whom?" is based on the "farmer-first" principle, which dictates that researchers begin with farmers' "knowledge, problems, analysis and priorities" (Chambers et al., 1989, p. xix). Although the specific term is mainly associated with agriculture in the nonindustrialized world, the idea has been advocated for the United States, too.

The "farmer-first" principle presupposes that farmers are the only beneficiaries of agricultural research; it cannot resolve conflicts between the interests of farmers and of other groups that support public research and are affected by it. Two of its advocates have suggested that the way to measure the value of research is the extent to which farmers adopt its results (Sumberg and Okali, 1989).

This narrow standard has been proposed for all agricultural research. It is thus an example of why we think that research about farming *methods* should be distinguished from research about agriculturally significant *processes* (chap. 4). The "farmer-first" view is more legitimate in the methods domain, but even there it offers only a partial standard. It does not take account of the desirability of a system for society as a whole, aside from its attractiveness to farmers and its technical feasibility. For example, a standard of worth based strictly on farmer adoption will give little credit to research on soil conservation techniques that are environmentally beneficial but expensive for farmers. Also, such a standard is not even meaningful, let alone appropriate, for research aimed not at immediate applications but instead at better understanding that eventually might yield improved techniques.

The "farmer-first" principle notwithstanding, increased influence for farmers in setting research priorities should not mean excluding other legitimate interests. Fortunately, the choice is not "farmer-first" versus "farmer-not-at-all." Allowing farmers a role in shaping research does not imply that other interests must be subordinated.

Conversely, researchers' obligations to other constituencies do not rule out participation by farmers in setting priorities. Instead, their involvement should be extended beyond their particular short-term needs. Many farmers are interested in the same issues as environmentalists, safe food advocates, rural activists, and so forth. Addressing farmers' long-run concerns presumably includes serious attention to such issues, too. It does not matter that farmers may not see them the same way as the other groups. Researchers can acknowledge farmers as "first among equals," a group with a special knowledge and feeling for agriculture but still only one of many "outside" constituencies whose views deserve to be heard, especially on matters that extend beyond the production system.

Farmer-Researcher Relations

Even if farmers' needs were the only ones that the agricultural research system was supposed to address, there would remain the question of how best to address those needs. Specifically, who is to decide what research questions should be studied?

The answer is both a practical and political matter. First, it affects the prevalent style of research. Research undertaken in response to farmers' demands tends to be shorter-term and more highly applied (Anderson and Lockeretz, 1991). Typically it deals with specifics ("What will work on my farm?"), whereas researcher-originated projects are more likely to strive for generality ("What do the results on this farm tell us about other farms?").

POLITICAL SIGNIFICANCE FOR FARMERS

Beyond this practical difference, the question of who should choose research topics is fraught with political implications for both farmers and research institutions. Farmers' lack of control over the research agenda is an especially touchy issue for alternative agriculture. Alternative agriculture largely arose as a conscious protest by outsiders, including some farmers, against the research priorities of the agricultural mainstream. No doubt it is galling to lose control over a new approach after struggling successfully to get it accepted.

This may explain why promotion of alternative agriculture often includes a call for greater farmer influence over research: both reflect outside groups' hopes of restoring an agrarian orientation to public agricultural research. To do so, they must counter the forces that render farmers less autonomous and more subordinate to external interests.

One such force is the economic power of input manufacturers. Research that advances alternative agriculture could reduce that power, because alternative agriculture usually connotes "low-input" agriculture, among other characteristics. Unlike the more limited notion of "reduced-chemical" agriculture, "low-input" implies that farmers try to become less dependent on suppliers of any input, apart from whether the input itself is undesirable.

A second force criticized for reducing farmers' autonomy is a research system regarded as having lost touch with the needs of its farmer clients. This gives more play to groups such as input producers. These groups can dominate the choice of research topics, either directly, by financing research, or indirectly, by their political leverage. The research done in response to this influence not only might fail to meet farmers' needs, but might even work against some farmers' interests. For example, it could result in technologies that force some farmers out of business by giving an advantage to larger and more heavily capitalized farms.

POLITICAL SIGNIFICANCE FOR RESEARCH INSTITUTIONS

The previous discussion obviously has implications for public research institutions, too. Advocates of farmer-initiated research who are outside the research establishment (e.g., Watkins, 1990) sometimes accuse those within its gates of an elitist unwillingness to share the prerogative of deciding what research should be done. Insiders, meanwhile, fear that uninformed "outsiders" will impinge on their academic freedom by demanding impractical changes in priorities and research methods (Jaschik, 1991). This fear is especially acute regarding new groups, such as "externalities/alternatives" organizations (Hadwiger, 1982, chap. 8). Unlike commodity groups and other older organizations, these groups sometimes take an adversarial posture, and they have not yet established comfortable working relationships with the administration and faculty of the land-grant universities.

Researchers sometimes have other reasons for not being sympathetic to farmers' demands for greater control over research. They may believe that existing mechanisms, such as a farmer advisory committee, are adequate forums for farmer-to-researcher and farmer-to-administrator communication. Furthermore, researchers may consider that they themselves are not particularly powerful in setting research directions; institutional administrators, budget limitations, and state and federal legislation all influence what researchers can do.

Allowing farmers to initiate research must be seen in this political context. That is why it is promoted as a tool for making policy changes, not just as a matter of research strategy. For example, for one group of farmers who initiated research on their own farms—an effective way for farmers to get research on a particular topic, but not the only way—its "greatest impact may have been its dramatic influence on the agricultural establishment" (Krome, 1988a, p. 6). The university responded to "growing demand from farm and rural advocacy groups for significant input into the policies of university programs . . . [by] creating research, extension, and curriculum programs that directly involve farmers and rural citizens" (Stevenson and Klemme, 1991, p. 1).

Similarly, a characteristic of the "participatory research" ideal promoted by one stewardship advocacy organization is that it led to "agricultural researchers spending more time listening to farmers' problems" (Nopar, 1990). This ideal is shared by some researchers in land-grant institutions: an example is an "active participation approach" in which "farmers can be full

125

partners in the development of new research-based information and recommendations" (Francis et al., 1990, p. 158). Whatever mechanism is used for researchers to learn about farmers' priorities, if it is a continuing process, with ideas and information going in both directions, an additional advantage is that farmers will improve their understanding of research and can then make even more valuable recommendations.

The Complementary Roles of Farmers and Researchers

The examples cited at the beginning of this chapter illustrate the view that farmers should have the dominant role in planning and evaluating research. This role includes not only setting general priorities but also influencing individual projects: the topic should be chosen according to farmers' preferences, and the project's value should be measured mainly by farmers' answer to the question "How useful was the work for you?"

However, farmers might not give high priority to agriculture's environmental effects, off-farm socioeconomic consequences, or long-term resource implications. Does this mean we should ignore these matters?

It is understandable if farmers give priority to research on their immediate problems. They are in a difficult and uncertain business; it hardly is fair to expect them to care more about problems that might affect other people some time in the future than about problems that already weigh heavily on their own farms. But empathy with farmers' preferences is not enough reason to make them the overriding consideration in designing a research program, whether alternatively oriented or otherwise.

Defining the purpose of research as meeting farmers' expressed needs, and interpreting this to require their approval of individual projects, may shut out legitimate modes of research. A well-planned research program requires a mix of projects: long-term and short-term; component and system-level; concerned with specific production systems and with general processes; and local, regional, and national.

Researchers, farmers, and other groups affected by research look at the same question differently, and properly so (Molnar et al., 1992). It is not surprising that some farmers suspect careerism when they see researchers doing work that lacks apparent usefulness. (Such suspicions are not always unfounded, as we discuss in chap. 11.) But from the researchers' viewpoint, a project should seek more than a self-contained answer to a specific question. It also should add to a cumulative body of knowledge, filling the most important gaps that prevent us from developing better agricultural systems. Thus the role of researchers is not limited to providing the technical expertise needed to answer questions posed by individual "clients." Do we really want researchers who think it sufficient to say, "Just tell us what should be done, and we will do it"? If so, who is the "you" to whom researchers should be so obliging?

Such an attitude has gotten us into trouble in the past. Agricultural researchers have been too ready to subordinate broader social considerations to the short-term technical needs of specific clients, such as commodity groups (Busch and Lacy, 1983, pp. 26–28). Excessively specialized farms and overuse of chemicals—two problems that alternative agriculture hopes to correct—may have occurred not because the land-grant system has been "elitist and exclusionary" toward farmers in general (Watkins, 1990, p. 161) but because it has been "too responsive" to the most influential farmers (Gardner, 1990, p. 171). The system's early orientation toward the immediate needs of farmers with the most political clout yielded indisputable benefits, but it also had its costs. Valid longer-term national goals, such as nutrition and food safety, were ignored or even opposed (Hadwiger, 1984). The system became

> accountable only to its users and, in other ways, unacceptable by the standards of today's science. . . . The system was isolated from the larger science community, and its training was for the most part narrow. . . . There was no accountability to the state as a whole. . . . The system depended upon artifice in disguising the selective benefits to clients. (Hadwiger, 1984, p. 197)

In the past, the research system's preferred clients were a minority of especially influential farmers, usually those with large, highly capitalized farms. The call now is to meet the expressed needs of another group, the more deserving "alternative" farmers. The point is the same, however: research has something to offer beyond meeting the immediate technical needs of any single group.

When researchers welcome farmer involvement in research, they are reversing an ill-advised exclusivity that has condemned some research areas to sterility. But when they carry the point too far, they are abnegating their professional responsibility. To have one group call all the shots is clearly not the key to a socially responsive research system. It is wrong for researchers to claim the right exclusively, and it is just as wrong to confer this right on farmers, whether "alternative" or any other kind. Farmers and researchers have different purposes, different kinds of experience, and different relationships to agriculture. Neither can do the job alone, but if they are allowed to complement each other, each can make a valuable contribution.

Part Three

BRINGING OUT THE BEST
IN AGRICULTURAL RESEARCH

10

· · ·

Putting Each Kind of Research in Its Place

Like the All-Seeing One, the hundred-eyed Argus of antiquity, like Briareus of the hundred hands . . . [the experiment stations] have suffered nothing to escape their close scrutiny and inquiry. From the pure rain drops of heaven to the drainage waters of the earth, and from the capture and imprisonment of the free nitrogen of the atmosphere to the composition, utilization and value of town sewage, —they question them all. . . . Whether they answer in the tongue of the chemist, the botanist or the physiologist, the answer has invariably been in the direct interests of practical, progressive agriculture.—Goodell, 1899, p. 27

Agricultural research entails many styles of work. Each can make a valuable contribution, provided it is used only where appropriate, without displacing other styles that are equally legitimate for other research purposes.

The institutional setting in which research takes place and the policies that govern it should allow each style of research to make its maximum contribution. In part 3, we suggest reforms to achieve this goal. In chapter 11, for example, we suggest diversifying the methods of allocating professional rewards so that they become compatible with the full range of legitimate research approaches. In chapter 12, we advocate a reasonable level of flexibility in grant program guidelines. Similarly, in the final chapter we discuss the advantages of attracting a more diversified student body and preparing students for varied research styles.

This chapter suggests a change in how we divide the territory covered by agricultural research. We believe that current disciplinary and departmental divisions impede adoption of the worthwhile innovative approaches we have been discussing, and that agricultural research can be made more effective if each style of research is housed in a more compatible institutional home.

The proposed change has two steps. First, some areas of research should be more clearly distinguished, so that they do not interfere with each other. Second, they can be grouped into new combinations, so that people using similar approaches and working on related problems are more likely to be part of the same department or program.

Ways of Dividing Agricultural Research

The institutional divisions of agricultural research, such as academic departments, are based mainly on the organisms or processes being investigated. Thus any research about insects can be considered entomology. Within an entomology department, one faculty member might investigate how different tillage systems affect the abundance of some pest, while another classifies insect species. Likewise, the money side of farming is the concern of the agricultural economics department, one of whose members might study the profitability of various pest control methods while another deals with world supply-demand-price relationships. Similarly, one member of an agronomy and soil science department might study the effects of tillage on soil properties while another classifies soils.

An alternative way to divide agricultural research would be based on whether or not it is closely linked to farms and production methods. In this scheme, one kind of department would deal with agriculture as a production system and would cover topics like the first member of each pair just described. Any study involving yields, production costs, environmental effects of production methods, or farmers' attitudes, for example, would fall in this category.

Another kind of department also would study agriculturally significant processes or organisms, but abstracted from the context of a production system. The reason for both kinds of research is the same: to improve our agriculture. In style, however, the second kind is less characteristically agricultural and more like other branches of science.

If we also were dealing with basic research, it would belong to a third kind of department in this scheme. However, we have excluded basic research from agricultural research (chap. 4). Even if basic research investigates an agriculturally significant organism, its value for agriculture, if any, will be indirect and unpredictable. Therefore, the division of research that we are suggesting involves only the first two kinds, both of which we consider applied.

We think that a sharper separation should be made between these two kinds of research, which now are often part of the same department. Once that is done, the portions of each department's work that involve production systems could be consolidated into a smaller number of production-oriented departments. The suggested separation would make all departments more homogeneous in the kinds of research questions they address. On the other hand, the subsequent consolidation would make some departments broader in the range of phenomena they cover.

The consolidation would be a shift toward multidisciplinarity, but it would affect mainly research on production systems. Under the current structure, the entomologist who studies how tillage affects pest abundance probably works in the same department as the entomologist who classifies insect species, but a more appropriate institutional home might be with the agronomist who deals with how tillage affects soil properties. On the other hand, the current disciplinary arrangement often is appropriate for research that studies processes abstracted from production systems (e.g., the entomologist who classifies insect species).

Separating Research Areas Qualitatively

WHY SEPARATING SOME COMPONENTS CAN BE DESIRABLE

Separating research areas according to whether they deal directly with production systems can have two benefits. First, topics can be regrouped into more fruitful combinations, as just illustrated. Second, the separation can reduce conflict between researchers who favor different approaches. Splitting a department is not the only way to achieve this goal. It also can be accomplished by formally recognizing different areas of research within a department instead of leaving them to coexist in an unspecified way.

Whichever degree of differentiation is adopted, the object is to allow different kinds of investigations to flourish by eliminating needlessly restrictive ideas about the "right" way to do research. Problems arise when the prevailing tone of a department or program is set by people who do not acknowledge the full range of legitimate approaches. For example, grant programs concerned with specific topics sometimes assume that projects necessarily will be linked to farms and farmers. The results can be incongruous. Thus one program gave "highest priority . . . to projects where farmers actively participate in the design and implementation" (National Research and Education Program on Sustainable Agriculture, 1991, p. 8); this priority applied even to research involving an "examination of a closely delimited biological or physical science problem . . . [such as] improving the nitrogen-fixing capability of a legume" (p. 5).

The idea behind our suggested separation is to avoid the "chauvinistic divisiveness" and "destructive competition" that beset agricultural research (Johnson, 1984). It does not seem enough simply to hope, as does Johnson, that a "covenant" of some kind might emerge on its own between researchers oriented toward production systems and those concerned with more general kinds of agricultural knowledge.

Making this happen might require some institutional realignments. Specifically, it could be fostered by separating areas that now must compete for funds, faculty positions, and status. Although such a separation might impede intellectual exchange among researchers, a different outcome also is possible: constructive intellectual exchange could be fostered by the reduction in administrative tensions and professional rivalries. Let us hope that intellectual exchange does not require that researchers belong to the same department; if it does, agricultural research is in serious trouble.

For practical reasons, we need some system for dividing agricultural research into areas of manageable size. Any such system will separate some researchers from others who do related work. As long as we must draw dividing lines, why not draw them in a way that separates people with conflicting views about valid research approaches?

A similar point has been made regarding the relationship between applied and basic research. Vollmer (1972) has argued that they should be *insulated* from each other, but not *isolated*. Their norms are different, and neither should be allowed to trample on the other. Instead, they can and should learn from each other. (Although the distinction we are suggesting concerns two kinds of applied research, they lie along a dimension that extends to basic research, as we noted earlier; Vollmer's argument is just as applicable here.)

WHY THIS SPECIFIC SEPARATION?

Unproductive conflicts can occur both within and across current disciplines. Rivalries between disciplines are a well-known obstacle to multidisciplinary research. However, not all of the conflict between agricultural disciplines arises because they study different phenomena. Some of it has its roots in different styles of work—"soft" versus "hard" science, quantitative versus qualitative, theoretical versus empirical, and so forth. Thus, there can be severe conflicts between people who pursue different styles of work even within the same discipline.

On the other hand, conflicts seem less likely between researchers whose work is strongly linked to an agricultural production system, even if they are in different disciplines. For example, the agricultural economist who studies the profitability of pest control methods will probably have less difficulty getting that work accepted by an entomologist who has developed those pest control methods than by an agricultural economist who works with mathematical models of world supply-demand relationships. The same point applies to two scientists from separate disciplines who both work on questions not directly tied to agricultural production. Scientists who specialize, respectively, in classifying insects and classifying soils, say, have little basis for quarreling over whose research is more legitimate.

If we want to separate research areas to reduce conflicts, we must do so according to what research approaches each takes, not just which organisms or processes it studies. In part 2, we discussed several alternatives to traditional approaches: research that is multidisciplinary, that takes place on farmers' fields, or that gives farmers considerable control in choosing topics. These less traditional alternatives are not fully accepted in some agricultural research circles. For example, some researchers harbor an unjustified prejudice against on-farm research (chap. 8) or are afraid of yielding even partial control to farmers (chap. 9).

Whether a less traditional approach is suitable for a particular research question depends a good deal on whether the research concerns a production system. Typically, research about farms or farming methods is more likely to benefit from being done on farmers' fields. In contrast, research that is independent of production methods does not often require the use of farmers' fields—in fact, it might not be done on any field. On the other hand, the subject of the research—whether insects, weeds, soils, or whatever—by itself tells us little about which research approach is appropriate. Separating research qualitatively, by its relationship to farming systems, can avoid inappropriate constraints on how the work is done because it will tend to separate researchers who use different approaches.

Making this distinction clear also may attract a greater variety of future researchers into agriculture. Depending on the type of research they pursue, some students eventually will work "out in the world," that is, in close contact with farms and farmers. Some no doubt will relish this prospect. Others would be just as happy to avoid it, preferring a style that is more like that of the non-agricultural sciences, even though the subject is agricultural. The latter might be put off by blanket statements that do not acknowledge the distinction, such as the pronouncement that "it would help immensely if all agricultural researchers had some practical farming experience" (Thornley, 1990, p. 176).

HOW COMPLETE A SEPARATION IS DESIRABLE?

Averting unproductive competition and excessive restrictions on research approaches can be achieved through various degrees of separation. Some institutions will not wish to split their current departments. If so, they can formally differentiate the positions of people doing different kinds of work, giving explicit labels to two kinds of research that now are not distinguished. This already is the case for faculty members who have responsibilities for teaching or extension as well as for research. Similarly, in biomedical research it has been argued that medical schools should establish a "two-platoon" system that gives equal recognition to both clinicians and academic investigators (Petersdorf, 1986), a distinction similar to what we are proposing.

If a formal distinction is made between the two kinds of work (even if they remain in the same department), it could help ensure that each has an adequate institutional base. Their respective fortunes should not be vulnerable to a temporary shift in tastes or to a change in influence within the department. In judging a researcher's work, for example, or in developing guidelines for a grant program, one must allow for all kinds of agricultural research, not just one's own.

Identified subdepartments are already common in agriculture, although more often the basis of the division is subject matter. For example, agricultural economics and rural sociology sometimes are housed in the same department. Nevertheless, their respective identities are clear, and the legitimacy of both is beyond challenge. The members of one cannot seriously criticize the research done by members of the other.

Nonagricultural science offers examples of recognized departmental subdivisions that are even more like what we are suggesting, being based on how the research is done, not on what gets studied. For example, experimental and theoretical physicists study the same phenomena and clearly belong to the same discipline. Yet the two groups have almost no overlap; only rarely is an experimentalist also a theorist. Moreover, the two kinds of physicists often do not even collaborate; although their respective bodies of knowledge are inseparable, individual research projects often are entirely theoretical or entirely experimental. Therefore a department in principle could function with only one or the other kind of physicist. A respectable physics department nevertheless must be strong in both because each is so well established that its absence would be glaring. Likewise, neither a theorist nor an experimentalist can be slighted for not being the other.

Combining Areas from Different Departments

If departments are divided according to whether or not they are connected with agricultural production systems, portions of several departments' territories could be merged into new combinations. This could be especially fruitful for topics that are closely tied to farms and production systems.

For example, in discussing multidisciplinary research (chap. 5), we described how an agricultural economist, a rural sociologist, and a soil scientist all might be interested in ways to conserve soil. This common interest could connect them more strongly to each other than to their disciplinary counterparts whose work is not linked to production systems, such as a soil scientist who specializes in soil classification, or an agricultural economist who deals with world commodity markets.

THE PRINCIPLE BEHIND THE REORGANIZATION

In the new arrangement, some departments would study farming methods broadly, linking their work strongly to the farmers who use them and the farms where they are used. The core of these departments would be the existing departments that deal with the major classes of agricultural products, such as horticulture, animal science, and agronomy (in its narrower sense of field crop production, but not including soil science).

These departments already tend to be multidisciplinary, but their scope can be broadened by absorbing research areas from other departments. Many disciplines, such as entomology, agricultural engineering, and plant pathology, could contribute to the new departments. The social sciences, which usually are completely separate, could make a particularly valuable contribution. In addition, some subjects could be moved out of these departments—for example, some biochemistry and plant physiology that now is housed in an agronomy or horticulture department.

Bringing together people from different disciplines who have a common interest in a major class of production systems is similar to the idea behind farming systems research. However, not every study in such a proposed department must deal comprehensively with a complete farming system, or have the "farmer-first" orientation of farming systems research. Nor must every study be multidisciplinary, or be confined to farmers' fields. An agronomist in the broadened version of an agronomy department might, for example, compare organic and inorganic fertilizer materials in much the same way as they now are studied in agronomy departments.

The difference would be that the agronomist's colleagues also would be interested in this question because of a common interest in many aspects of farming methods. They could help interpret this research in the broader context of an entire production system.

The presence of such people nearby does not by itself insure that collaboration will happen, of course; one can go for years without interacting significantly with a departmental colleague in the next office. Presumably, though, researchers who choose to work in this new arrangement will want such interaction. Compared with an agronomist working in a traditional agronomy department, therefore, the agronomist in the proposed department probably would give more thought to related questions: the economic significance of the results; farmers' attitudes toward different kinds of fertilizers; and the possible effects of type of fertilizer on soil pathogens, weeds, or insect pests.

CHOICES IN HOW DEPARTMENTS ARE REORGANIZED

We cannot give a simple blueprint for these new combinations. Whether a particular school should change will depend on its size, the variety of agricultural systems it deals with, and its strengths in various agricultural and nonagricultural disciplines. For example, a large college in a state with a diverse agriculture might choose to have several departments dealing with production systems. A small college, in contrast, might go with only one, called simply "agriculture." It might seem strange to have one department called "agriculture" in a whole college devoted to agriculture. Yet the same situation prevails in medicine; despite the growing importance of medical specialization, the department called simply "medicine" is still the heart of the medical school.

The suggested rearrangement may seem unnecessary in a large university, where faculty members from various departments perhaps can collaborate now under arrangements that supplement rather than realign existing departments. These units, variously called "centers," "programs," and so forth, are suitable for doing research on particular subjects, but they are not necessarily satisfactory for educational purposes. Even if a nondepartmental unit can grant a degree, it might be restricted to courses already offered by the departments, offering none of its own design. This limits the opportunities for reforming graduate curricula; in chapter 13, we discuss some possibilities that could be aided by the departmental realignment we are discussing.

In a small university, by contrast, members of different departments already may be in close enough contact to work informally in the way that the new alignment is intended to encourage. Realignment might not be feasible anyway because the faculty may not be deep enough. Out of necessity, the school might not be trying to cover the entire territory of agricultural research. Under these circumstances, our ideas about how to divide the territory more effectively are irrelevant.

Issues in Departmental Reorganization

HOW FIXED ARE DEPARTMENTAL BOUNDARIES?

Departmental reorganization is a difficult and sometimes painful process. People who are accustomed to a particular departmental structure often are reluctant to change it. Still, one should not assume that current departmental and disciplinary boundaries are immutable.

Dairy science is an example of a discipline that arose by combining areas from several older disciplines. Departments of dairy science exist because a common interest in the dairy cow was more important for some researchers than their ties to the traditional disciplines in which they were trained, such as genetics or physiology. The changes we suggest would carry such a transformation further by consolidating an even broader range of research topics.

Another reason to think that departmental realignment is feasible is that among land-grant universities there is no uniform set of departments. For example, plant physiology, plant pathology, entomology, and nematology have their own departments at some universities, but they occur in various combinations at others. Similarly, soil science, crop science, and agronomy may be covered in the same department or in separate departments.

DOES DIFFERENTIATION IMPLY FURTHER FRAGMENTATION? Making a sharper distinction between qualitatively different kinds of agricultural research could exacerbate a disciplinary fragmentation that already is excessive. On the other hand, under the suggested arrangement, diverse approaches would be more likely to flourish within the same institution (even if not in the same department). Each researcher would thus have opportunities for more varied kinds of collaboration.

Some institutions have been emulating basic sciences by pushing aside the more farmer-oriented research; conversely, a department that gives top priority to farmers' immediate needs will not be a very hospitable home for research that is not linked directly to farms and production methods. Our proposal is intended to keep both orientations strong by assuring that both will have recognized identities. The result will be that across the institution as a whole, a researcher will have a more diverse range of colleagues.

Still, we must acknowledge that under the proposed reorganization, practice-oriented areas of research would risk becoming stagnant through isolation—that is, they might not get refreshed with new scientific ideas. Conversely, researchers concerned with more general agricultural principles would risk losing close contact with researchers whose empirical observations in the field might suggest interesting scientific questions. (These problems occur even under the current structure, however; just being in the same department does not guarantee productive intellectual exchange among researchers, any more than being in different departments prevents it.) The risk can be reduced by formal mechanisms for interchange between the two kinds of departments we are proposing, like the grand rounds that bring together people from various departments of a medical school.

Also, the different orientations can be bridged by suitably motivated individuals. The mechanism would be analogous to programs for "physician-scientists," clinicians who receive supplemental training in a basic biomedical science and thereby bring a broader perspective to clinical research.

THE PROBLEM OF DIFFERENTIAL STATUS

Another possible objection to the suggested differentiation is that it may leave the more farm-oriented kind of research with second-class status. Such stratification already occurs informally in some circles, even though the two kinds of work do not have formal identities. Alternative sources of prestige may become more important, however, thereby shifting the balance.

Traditionally, the more applied kinds of research have been looked upon less favorably in many scientific disciplines, including the agricultural sciences. This point will be central to our discussion of the professional reward system (chap. 11). But when a field has high visibility outside the academic world, prestige need not come only from other researchers. Do engineers have lower prestige than physicists (other than among physicists)? Do M.D.s doing clinical research have less prestige than cell biologists (other than at institutions where the latter set the tone)? To put it in the crassest material terms, what does it say about comparative status that an M.D. at the National Institutes of Health (NIH) can receive a pay differential above the scale for a Ph.D. with comparable experience? Similarly, the prospect of a "breakthrough," especially for a disease that evokes a strong emotional response, can elicit philanthropic support much greater than research grants that basic scientists can hope to get from the highly competitive NIH study sections. One can think whatever one wishes about the comparative intellectual merits of these two kinds of work. However, it is hard to argue that a clinical researcher in charge of a fully funded research unit has a problem of status compared with a basic researcher who must scramble for each new piece of equipment.

In agricultural research, status also can come from the backing of a constituency that feels it has received important benefits. Granted, such backing can be bought too dearly. The researcher may become beholden to the most influential groups, whether agribusinesses, environmental activists, farmers' organizations, or any others who would like to influence the direction of research: "The problem of maintaining effective communication with outside groups without becoming either a captive or an adversary is exceedingly difficult" (Ruttan, 1978, p. 6). Granted, too, high regard from constituents is different from the purely collegial status that traditionally has been prized. But status among one's colleagues is not the only thing that determines how well a professional fares.

We are not implying that the opinions of other professionals no longer matter. However, each kind of work should be judged according to its own goals. Unfortunately, in a department dominated by people with a particular orientation, researchers with a different orientation may feel they must try to pass their work off as something it was not intended to be. The proposed departmental realignment would help rectify this by making each department more homogeneous in its goals and in the research approaches it uses to achieve them.

A POSSIBLE OBJECTION TO NEW COMBINATIONS

A legitimate concern raised by broadening the scope of some departments is that it could encourage superficiality by sacrificing depth and forgoing disciplinary control over quality. We offer three counterarguments.

First, the proposed arrangement would not replace the existing system of disciplines. It would only modify it, by broadening the departments that do research on production systems. Other departments would continue doing the disciplinary research for which they are best suited.

Second, not every member of such a department would be expected to be knowledgeable about every discipline contributing to it. Such a requirement indeed would lead to superficiality. Instead, department members can be people who are trained in a traditional discipline but who prefer working with a broader range of colleagues. (This differs from multidisciplinary research as described in chapter 5 in that the interactions among members would be longer-lasting but less formal and less close than among collaborators on a specific project.) There is good precedent for such an arrangement. For example, some people trained in a service-type discipline, such as statistics, choose to participate in agricultural research rather than developing improved statistical techniques. Their disciplinary competence is not impugned by that choice.

Third, as we have already discussed (chap. 5), the notion of a "discipline" has two connotations. One concerns subject matter: how questions are chosen for study. The other refers to professional and institutional functions, such as maintaining quality. Regrouping the subject matter covered by various departments does not mean abandoning the professional aspect of disciplines.

Implications for Older Departments

The proposed realignment would mean that most departments acquire some areas but give up others. Those whose work is not tied to production systems would likely suffer a net loss. For example, some areas of plant pathology might be shifted to the broadened version of the agronomy department, without an offsetting shift into plant pathology.

A department would be willing to give up territory only if the new arrangement had some other attraction. One benefit is that the department could insulate itself more from short-term pressures to meet specific needs. Similarly, experiment station researchers were happy when the Smith-Lever Act relieved them of extension responsibilities, thereby allowing them to do more scientific research.

With the most farming-related parts of a department's territory moved elsewhere, the research that remains would be closer in style to that of nonagricultural departments. The main difference would be that it still involved organisms and processes that are agriculturally significant. Scientists who prefer such work would be justified in pursuing it without laying themselves open to the charge that they have ignored their social responsibility to improve agriculture. Rather, they would be fulfilling it in one way, while their colleagues in the production-oriented departments were doing so in a more immediate and direct way. Diverse missions among departments should be tolerated because what counts is how each department contributes to the university's overall mission (Schuh, 1984).

Having given up their most production-related work, departments such as plant pathology could continue as independent units, or they could merge with their nonagricultural counterparts. The latter choice would allow closer contact between basic scientists and those agricultural researchers for whom such contact would be especially valuable, as recommended by the Pound Report (National Research Council, 1972).

The suggested rearrangements could solve a dilemma that many agricultural science departments face because of conflicting demands. On one hand, they are expected to meet specific agricultural needs, which means doing research in a way that is characteristically agricultural. On the other hand, they are aware that they cannot ignore developments in other scientific fields (for example, in molecular biology), which pull them away from the traditional domain of agricultural research.

Caught in this dilemma, striking a balance within one department is difficult; typically a given department will respond more in one direction than the other. As an alternative, we should consider consolidating the most farming-related parts of each department's territory into a few broad departments that are especially committed to such work, while tying the remaining parts more closely to the corresponding general science discipline.

11

. . .

The Professional Reward System

There was a meeting of agricultural economists at which the usual things were discussed—the effect of the gold supply on trade, the relationship between the abstraction "agriculture" and the abstraction "industry," the influence of the price of hogs on the delivery of hogs to market, the preference of potato buyers for clean over dirty potatoes. Then suddenly a delegate arose and remarked that agricultural economics was an applied science. . . . What kind of rural life, what kind of rural society, do the scientists concerned wish to produce, if any? . . . This forthright approach rather threw the meeting into consternation.—Harding, 1940, pp. 1,100–101

The problems in getting alternative approaches accepted are reinforced by the customary way of rewarding professional achievement at institutions that follow traditional academic criteria (universities, mainly, but also some government research facilities). That is why alternative agriculture advocates often talk of the need to change the ways of deciding matters such as tenure and promotion (Soule and Piper, 1992, pp. 66–68, 212–15).

However, reforming the academic reward system involves more than removing the obstacles to particular kinds of research. Therefore, after briefly describing those obstacles, we consider the reward system in general, from its origins in the pure science disciplines that emerged in the nineteenth century to its later application—and misapplication—to other fields, including agriculture. Believing that the system often conflicts with the social purpose of agricultural research, we cite several grounds on which it can be challenged. Although individual researchers can adapt to the dominant system while pursuing alternative approaches, strategies to do so are too limited. What is needed is a professional reward system that is compatible with the diversity of worthwhile agricultural research.

Research in Relation to Other Professional Activities

This chapter deals only with rewards for research. However, research is just part of the job for many academic researchers. Therefore it must be given appropriate weight relative to other professional activities. Inspired teaching, dedicated service to the department or university, excellence in writing or speaking to nontechnical audiences, effective extension programs, and so forth—all deserve to be recognized and rewarded when they are appropriate for a particular position.

Unfortunately, even where these activities are supposed to be important, the prestige of research may overwhelm all other work. To avoid this, the relative merit ascribed to various activities should be made clear when the researcher is hired. It may be awkward to do this. A department that rhetorically attaches high value to creative teaching, say, would prefer to leave unsaid that research achievements count for more.

A balanced evaluation of a candidate for a position that requires more than research should not, however, overcorrect a tendency to downplay nonresearch activities. Good presentations to farmers' groups, for example, while worthy in their own right, do not compensate for a mediocre research record if research is understood to be an important part of the job. The following discussion presupposes that when a researcher's overall professional performance is being evaluated, research gets whatever weight is appropriate for the particular position.

Problems in Pursuing Innovative Research Approaches

In discussions of the professional obstacles to certain research modes, such as multidisciplinary research (for example, MacRae et al., 1989), the criticism heard most often is that researchers must produce many papers, which must be published in the leading journals of the discipline. Also, substantial credit goes only to a sole author or, for a multiauthor paper, to the first few authors.

When these criteria are applied rigidly, they work against research that involves several disciplines or that yields publishable results only after several years. Also, for investigating an unfamiliar "alternative" system, an exploratory study may be most appropriate, but some disciplinary journals, especially in the social sciences, require that papers propose or test a hypothesis (Berk, 1981).

A similar problem can arise with on-farm research. Because it may not use familiar experiment station methods, reviewers for traditional disciplinary journals may be prejudiced against it (Anderson and Lockeretz, 1991). Unfortunately, the proper criterion—"Does the paper use the right methods for its purpose?"—may degenerate into "Does the paper use 'correct' methods?" In chapter 4, we noted that a prejudice in favor of hypothesis-testing statistics not only obstructs publication of valid research but also encourages researchers to perform, and journals to accept, analyses that are wrong despite the use of standard techniques.

It is difficult to determine how strongly the reward system constrains what researchers can do. The criticisms may be exaggerated because the issue may be raised mainly by people who have not been well served by the dominant system. Perhaps we hear less from people who have been successful despite their less traditional research approaches. Also, some disciplines and institutions handle these matters differently from what we call the "dominant" system. Therefore the arguments of this chapter do not apply everywhere. Still, the distorting effects of the professional reward system deserve serious consideration, along with the possible alternatives.

Looking beyond Specific Kinds of Research

One approach to the problems described is to tailor the reward system to the special characteristics of alternative agriculture research, such as multidisciplinarity and collaborations with farmers (MacRae et al., 1989). However, such a specific change does not take account of the institution-wide constraints on what a particular department can do. As an example of more general procedural reforms, some universities now allow only a few publications to be considered in promotion or tenure proceedings, so that quantity cannot substitute for quality (Stetten, 1986; Mooney, 1991).

Such reforms are helpful, but we suggest a more aggressive strategy. Researchers who favor change should think beyond getting the system to accommodate their particular kind of research, and beyond procedural improvements that would let the current system achieve its goals better. Instead, they should challenge the very idea of the dominant system as poorly suited to the social purposes of agricultural research. (We do not, however, favor an approach that some departments are said to use [Beattie and Watts, 1987]: following alternative reward criteria in practice but disguising them by using traditional rhetoric to appease administrators. The practical consequences might be the same, but an alternative reward system should be instituted openly, not through subterfuge.)

We believe that the academic reward system should be questioned on two grounds. First, it was intended to serve only certain kinds of scholarship, and its underlying principles are not appropriate for many fields to which it later was applied, including agriculture. Second, even though it might once have been suitable in the ideal, professional and institutional conditions have changed greatly since the ideal developed. Despite the rhetoric, alternative reward systems already exist, partly as a necessary adaptation to these changes. Some alternatives are important mainly in fields other than agriculture. Nevertheless, even changes in those fields can support change within agriculture because the reward system cuts across disciplinary lines: researchers in one field are influenced by their perceptions of what is happening elsewhere.

In developing these points, our premise is that the purpose of the reward system in a publicly supported profession is to help ensure that the profession achieves the broad, long-term goals that the public had in mind in supporting it. Perhaps not all researchers will agree. Some might reject the notion of public accountability, preferring that the public supply the means and let the profession select the ends. Others might demand even closer accountability than is suggested by the words "broad" and "long-term."

Our position is intermediate. We assume that deciding how rewards should be allocated, and the process of allocating them, will remain in the hands of the profession. However, the profession should not carry out this function under the pretense that agricultural research is an autonomous, self-governing activity insulated from the rest of society.

The Principles Underlying the Professional Reward System

Proposals for changes in the professional reward system should start with the ideal on which it rests and with its historical evolution. The system is strongly rooted in the disciplinary organization of research. A key aspect of the original concept of a discipline is that the problems considered suitable for research arise from the discipline itself. Indeed, the purpose of research is to advance the discipline's knowledge. Traditionally, therefore, a discipline is intellectually self-contained:

> Our society is mission-oriented. Its mission is resolution of problems. . . . Since these problems are not generated within any single intellectual discipline, their resolution is not to be found within a single discipline. . . . The university by contrast is discipline-oriented. . . . The problems it deals with are, by and large, problems generated and solved within the disciplines themselves. (Weinberg, 1967, p. 145)

The reward system reflects this notion:

> Society's standards of achievement are set pragmatically: what works is excellent. . . . [In contrast] the university's standards of excellence are set by and within the disciplines. What deepens our understanding of a discipline is excellent. (Weinberg, 1967, p. 145)

Therefore the "products" of research are publications, especially publications in one's own discipline, and researchers in that discipline ("peer reviewers") are the best judges of their value. For disciplines that still fit the traditional concept, this standard makes sense.

The disciplinary basis of academic rewards has a long historical precedent, going back to the emergence of the research university in the late nineteenth century:

> One can, I think, mark the beginning of modern academic science in the United States at that moment when American investigators began to care more for the approval and esteem of their disciplinary colleagues than they did for the general standards of success in the society which surrounded them. . . . The decades between 1850 and 1880 saw Americans in a number of the sciences beginning to accept these values. . . . Success, in this world, meant acceptance as a creative scholar by one's disciplinary peers. Concretely, this demanded the publication of books and articles. (Rosenberg, 1966, pp. 154–55)

Emphasis on quantity of publications, which now is a particular target of criticism, became pronounced in the post–World War II period. It reflected the rapid expansion of scientific research, the increased specialization that accompanied it, and the sharp rise in federal funding of research through project grants (Kaplan, 1964).

Extension to Other Disciplines

THE PLACE OF PROBLEM-ORIENTED DISCIPLINES
IN THE UNIVERSITY

Although research universities at first dealt only with "pure" scholarship, today's universities cover a much broader territory. Many fields now considered disciplines in the looser, more modern sense, including the "quasi-disciplines" or problem-oriented fields (chap. 5), would not have been accepted at the early research universities.

Fields that produced mere practitioners, such as medicine and law, at first did not fit in, although they had been important in the classical university since medieval times. Nor did "agriculture and the mechanic arts," the heart of the land-grant college idea that was such a striking challenge to mid-nineteenth century notions of higher education. Even statistics and bacteriology, which were not linked to professions, trades, or industries, had to struggle for legitimacy in the research universities (Ben-David, 1971, pp. 142–52).

The obstacle for these fields was that they did not arise out of the internal dynamics of research but were a response to a need in the "outside" world. When research universities expanded to include the outsiders, "scientists in the established disciplines had many misgivings about the danger of blurring the borderlines between disciplinary science and problem-oriented research which often lacked theoretical significance" (Ben-David, 1971, p. 145).

INTERNALLY ORIENTED VERSUS PROBLEM-ORIENTED RESEARCH: AN EXAMPLE

This distinction is especially clear in research on the social and economic aspects of agriculture. Two rural sociologists, for example, might study why some farmers, but not others, participate in a voluntary soil conservation program. They probably would use similar tools, such as a mail survey. However, their reasons for studying the question could be different.

One researcher might use the answers to help the people in charge of conservation programs make them more attractive to farmers. The second investigator may see the same question as a way to test ideas that matter mainly to other people in the discipline. A paper with this orientation will be about a "model" of farmers' decision making, a "conceptual framework," a theory, or a hypothesis; that is, the study would justify itself using arguments internal to the discipline.

Both researchers would say they are working on a "problem." The difference is that the problem in the first study—farmers' nonparticipation in a socially desirable program—exists outside the academy, whether or not researchers choose to investigate it. In contrast, what the second study examines is a "problem" because people in the discipline have agreed it is (Johnson, 1971). The first kind of problem is a troublesome situation that we would like to improve; the second kind is a gap in our knowledge. Many fields, including agricultural sciences, deal at least in part with the first kind; strict exceptions, such as pure mathematics, where there is no "outside" world, account for only a small portion of research in the modern university.

But even in fields concerned with external problems, including agriculture, the reward system reflects the higher status of the second kind of study.

To share in the perceived prestige of "purer" kinds of research, these fields have embraced a value system that does not fit them.

The irony is that this value system once kept agriculture out of academia. As Ruttan (1991, p. 122) has put it: "The ideal of 'pure' science has been advanced to protect its privilege and its ego from contamination by engineering, agronomy, and medicine." He called for an end to "indefensible intellectual and class barriers" and "outmoded status symbols and cultural constraints" that keep "pure" scientists apart from those whose work has practical value (p. 122).

Yet these "outmoded status symbols" are perpetuated by some agricultural researchers when they criticize their colleagues for not advancing disciplinary knowledge as the pure sciences do. Perhaps such people should be reminded that a few generations ago—simply because they were agricultural researchers—they would have been on the receiving end of the same criticism (Danbom, 1992).

How an Inappropriate Reward System Distorts Agricultural Research

THE QUEST FOR PRESTIGE

The present reward system suits some agricultural research—mainly highly specialized work that is not tied to agriculture as a production system and a human activity. But the system is not appropriate for agriculture as a whole. The reasons go beyond the obstacles it creates for certain modes of research, such as studies that take too long to yield many papers per year. A more general consequence is the relative status of the two modes of agricultural research illustrated in the rural sociology example.

Orientation to problems in the outside world is the older mode, going back to the establishment of the land-grant university and experiment station systems. However, like many other fields that started with a problem-solving or service orientation, agricultural research has since come to emulate the standards of the "ivory tower" disciplines. The reason is that

> the social structure and the purpose of the university accent the pressure toward purity. . . . In the university it is improper to ask of the scientist, What is the relevance of what you are doing to the rest of the

world? . . . The acceptable question is, What do your scientific peers
. . . think of your work? (Weinberg, 1967, p. 150)

The norm of "purity" sometimes is taken even further, so that work tied to problems in the real world not only is unrewarded but may even be considered grounds for suspecting the researcher's legitimacy.

Perhaps this shift was inevitable, given that the research universities that established pure science research have been among those most prestigious in the country. Universities not known for strong research often copy the "pure" standards, regardless of suitability. They emphasize traditional notions of scholarship over areas where they might excel, such as high-quality teaching or research oriented toward local problems. For agricultural colleges in particular, this ill-advised approach has meant "denaturing the land grant tradition of problem solving and service . . . and turn[ing] good land grant universities into second-rate, private academies" (Bonnen, 1986, p. 1076).

True, the land-grant schools are proud of their nonelitist roots and their dedication to serving the public. They readily invoke the label of "the people's colleges," especially when the people's willingness to support them seems to be flagging. But agriculture has been eager to adopt the "look and feel" of fields that never had to prove that "we're real science too."

The perceived lower prestige of "applied" fields apparently has been accepted by some of those on whom its burden falls. Many agricultural disciplines, perhaps to escape the stigma of being labeled "applied," have taken on themselves a reward system intended for a different kind of research. Never mind that several of them originally separated from their "pure science" parent disciplines precisely so they could adopt disciplinary norms more appropriate for applied science (Rosenberg, 1971).

Agricultural research differs from the traditional "pure" sciences in its social purpose, and the reward system should be appropriate to that purpose. We are not suggesting, however, that it should remain isolated from other branches of science; it could benefit from more intellectual contact with many nonagricultural fields (Mayer and Mayer, 1974; Richards, 1988). Thus one observer, after criticizing the colleges of agriculture that are "retreating from the land grant vision," also chastised those that "build a moat around the college and withdraw from the university, shifting so far toward application that they are now isolated from basic science" (Bonnen, 1983, p. 963).

THE CONFLICT BETWEEN PRESTIGE AND SOCIAL UTILITY

Although disdain for the label "applied" is sometimes considered a recent development, it has been with us for decades, and it was strong enough by the 1930s to draw the vigorous criticism of sociologist Carl Taylor (see chap. 3). Taylor had organized pioneering applied research on rural communities in support of USDA's New Deal action programs. He also enjoyed an impeccable reputation in his discipline, as shown by his selection as president of both the Rural Sociological Society and the American Sociological Society. Taylor urged his colleagues to "abandon some of the pleasures . . . [of their] semi-esoteric ways of life" and participate in social programs (Taylor, 1947, p. 8). He believed their participation not only would help the programs but also would advance sociology as a scientific discipline by providing a good place to test and refine its theories (Taylor, 1941).

Despite this view, the conflict between a disciplinary orientation and the social purpose of fields like rural sociology has remained strong and has perhaps intensified. The resulting "decaying land grant vision" (Bonnen, 1983, p. 963) seems both unproductive and unnecessary. One cannot pretend that agricultural research is anything but an overtly "applied" field; when asking for public funds, agricultural researchers do not usually say that their main criterion in choosing a topic is what their disciplinary colleagues will think of it.

They do not even say that among themselves. In a study of how scientists select research problems, Busch and Lacy (1983, pp. 111–12) found that scientists in all twelve agricultural disciplines they surveyed rated "importance to society" higher than "scientific curiosity," "publication in professional journals," and "contribution to theory." In basic sciences the opposite was true.

WHAT IS WRONG WITH THE DOMINANT REWARD SYSTEM

In advocating a different reward system for agricultural research, there is no need to attack the prestige of the "pure" disciplines. Some people, justifiably decrying the applied disciplines' tendency to ape the pure sciences, carry the point too far: their broadside criticism of pure science sometimes seems ignorant and anti-intellectual.

It also is unnecessary to condemn the reward system of the older disciplines as wrong in an absolute sense. We focus on a more limited problem: the system simply is not suited to the goals of much agricultural research, and thus distorts research that otherwise might be valuable.

A researcher in an overtly applied area should have a clear and honest answer to the question, "What agricultural problem is your work trying to solve?" Because agricultural researchers are being pulled in different directions—by the demands of outside groups and by their need to be recognized by their disciplinary colleagues—they would like to count their work both as problem-solving and as contributing to improved disciplinary understanding.

Unfortunately, a research project often cannot meet both these purposes. Also, publication conventions might interfere. For example, the most effective contribution that social science researchers could make to solving some social problem might be an unabashedly descriptive or exploratory study. However, the major journals of the social sciences have little room for such research. Overwhelmingly, they emphasize advances in disciplinary understanding. To judge by their content, and sometimes also by their editorial guidelines, it seems that research does not need to lower itself to description. Instead, it can skip right to the "real" science—analyzing general social theories and testing hypotheses (Berk, 1981). (Where these theories are supposed to come from in the absence of descriptive and exploratory studies is not explained.)

As a result, a researcher who needs prestigious publications may transform the work into something intended to advance "conceptual understanding" of the subject, presenting it in the sterile formalism and jargon of the discipline. This is a sure prescription for having it ignored by anyone who could make real use of suitable work on this topic. Pressed to show the practical value of such work, the researcher might argue that the improved conceptual understanding eventually will lead to practical measures. However, this argument has been so overused that it should be viewed with extreme skepticism. The criteria for disciplinary publication say nothing about any "real-world" contribution.

Other authors, in contrast, will stick to an analysis and presentation style suited to their problem-solving purpose. This means that they risk not faring well professionally if publication in a top disciplinary journal is an across-the-board standard.

DISCIPLINARY PURITY: GROUNDS FOR SUSPICION?

In some branches of science, mainly outside agriculture, there is no conflict between a disciplinary or "pure science" orientation and the social purpose of the field. In pure mathematics research, for example, public funding is available even though an individual piece of research is not intended to have a direct and immediate application. Its tangible benefits come in the long term, if ever, and the researcher is not expected to anticipate them when the research is begun. In such a field, a reward system based on within-discipline considerations is entirely consistent with the field's social purpose, namely, the advancement of knowledge.

We feel obliged to say this because well-founded criticisms of the academic reward system sometimes are generalized into unfounded criticisms of certain kinds of research. Some people suspect that despite the rhetoric about social value, careerism is the real reason that agricultural researchers choose a style of work more like basic science (Bird, 1991, pp. 26–27). This cynicism ignores the diverse character of agricultural science. It would be unfortunate to assume that because a researcher's work is well rewarded under disciplinary criteria, it is not also fulfilling the purpose that makes society willing to pay for it (Thornley and Doyle, 1984).

Deviations from the Dominant System

To a degree that some people might not realize, the traditional reward system has lost its exclusive sway over many areas of academic research. The institutional conditions that prevailed when it was developed have changed greatly. In accepting these changes, the profession has rejected the premises underlying the traditional reward system without necessarily acknowledging the rejection.

There are several ways in which the "dominant" reward system does not dominate:

• Some scientific papers have so many authors that the usual procedures do not apply.

• Funding sources, not disciplinary colleagues, often determine what topics will be investigated.

• Much research is done outside the disciplinary structure.

• Financial rewards are important in some academic fields.

BIG SCIENCE AND MULTIPLE AUTHORSHIP

The growth of "big science" (Weinberg, 1967, pp. 106–8, 113–14; 1970) presents a serious challenge to the customary way of measuring a researcher's "productivity." No one doubts that big studies, such as those done at high-energy particle accelerators or radio telescopes, are "real" science. They do not fit the modes that prevailed in the early research universities, however. There, a scientist either worked alone or worked with assistants of much lower status, who did not share in the credit. In big science, authorship of single- or few-author publications no longer is a usable standard for judging a researcher's contribution. In experimental high-energy physics, for example, a paper can have several hundred authors.

Clearly, the system has had to adapt: if all that counted for credit was how many papers had one's name near the head of the author list, most high-energy physicists would never advance professionally. Author lists are so huge that it is out of the question to disregard contributions beyond the second author's, say. Instead, one must learn what the eighty-ninth author contributed to the project and then judge the significance and quality of that contribution. Why not apply the same principle to the much smaller research groups typical in agriculture, instead of giving credit only to the first author or two?

EXTERNAL FUNDING

The prevalence of external research funding presents another challenge to the ideal of the professional reward system. In accepting funding of specific research topics chosen by outside grantors, researchers are tacitly rejecting the notion that a discipline chooses its own research questions. Because funding is itself a kind of reward, this practice conflicts with the notion that rewards come from one's discipline.

This argument does not apply to project funding decided by peer review, such as the study sections of the National Institutes of Health. (Even there, political considerations sometimes override peer review of controversial grants.) But funding in agriculture and many other fields often is more targeted, even when it comes from a public source. For example, the source might be a mission-oriented agency or a legislature responding to specific political pressures, as we describe in the next chapter.

The conflict between external project-by-project funding and the ideals of the academic reward system arises most sharply in collaborations between public universities and private companies. In this increasingly common arrangement, the tone often is set mainly by the latter:

> With greater intermixing of public and private motives, the public sector . . . must find new ways to assure the integrity of science and its decision processes. Private purposes can easily dominate joint ventures, thus forfeiting much of the larger social benefit that might otherwise be achieved from collaboration between the public and private sector. (Bonnen, 1986, p. 1075)

Private external funding may lead to a flagrant violation of the very idea of academic research. For example, a researcher may have a financial involvement with a company that supports the research, and therefore stands to gain if the results turn out a particular way. Academic researchers are becoming increasingly distressed about such conflicts of interest (Marshall, 1990b).

The problems are not limited to abuses, however. By nature, industry-academic collaborations face a possible conflict with the traditional scientific norm of open access to data (Marshall, 1990a), and therefore with a reward system based on peer review:

> The academic responsibility of open communication inevitably conflicts with the commercial responsibility to maintain proprietary se-

crecy. . . . Secrecy in research precludes replication and weakens the
system of peer review. (Nelkin, 1984, p. 25)

Even where secrecy is not an issue, a fundamental point remains: industry-
university collaborations show that many researchers do not work in isolation from the larger society, hoping to win recognition only from a narrow
circle of colleagues, the only people who understand or care about their
work. That was the ideal of science and, to a substantial degree, the reality
when the discipline-based reward system began (Weinberg, 1970). Whatever one thinks of the ideal, it hardly fits most research today. Thus it is
ironic that some land-grant universities that "talk boldly about developing
partnerships with high tech industrial firms," also are "retreating . . . to the
ivory tower of disciplinary purity" (Bonnen, 1983, p. 963).

NONDEPARTMENTAL RESEARCH ORGANIZATIONS

The changing organization of research makes untenable the idea that contributions to one's own discipline are all that count. In agriculture as elsewhere, alternatives to traditional academic departments, whether called "institutes," "centers," or whatever, have become the institutional home for
much multidisciplinary research (Boger and Boyd, 1982). In agriculture,
these programs deal with issues ranging from the problems of small farmers
to commercialization of unconventional crops, to reduced-chemical farming
systems.

In the past, senior researchers in such organizations usually held conventional faculty positions; those who did not were considered second-class
compared with similarly qualified regular faculty (Teich, [1979] 1986). This
distinction no longer is generally true. Highly regarded nondepartmental research units may offer attractive positions that entail no appointment at all in
a regular department, or at most a courtesy appointment.

Discipline-oriented professional standards obviously do not fit research
units dealing with issues that are not defined in disciplinary terms. Such
units are important in today's university, even though they "suffer from a
mismatch [with] the social ethos of the university" (Weinberg, 1970, p.
1070). They are a break with the ideal of the largely autonomous academic

department that prevailed when the research university became important about a century ago.

Their importance, however, does not mean that more such units should be established just to get around the monopoly of the discipline-based reward system; a poorly conceived multidisciplinary center can sacrifice the strengths of a department while becoming just as rigid, even if the way it describes itself sounds fresh and innovative. Our point is only that the monopoly has already been broken. This precedent provides a powerful argument to people trying to introduce more flexibility into a reward system that is too narrow for the varied goals of today's agricultural research.

PROFESSIONAL AND PRACTICE-ORIENTED FIELDS

Another alternative to the traditional reward system occurs in professional and practice-oriented fields, such as engineering, medicine, and agriculture. We have already noted that older disciplines were prejudiced against these fields. Nevertheless, their practical value eventually outweighed the prejudices (Ben-David, 1971, p. 146).

Despite their academic acceptance, these fields remain linked to the outside world. In other words, the academy now includes many people who also have nonacademic rewards available to them, such as consulting contracts, royalties, and private practice. Not only has the academic world accepted such people; it also has been quick to let them combine academic positions with well-paying outside activities. Moreover, the prestige of their university appointments enhances their chances of being offered such opportunities. No doubt this would have shocked the leaders of the pristine research universities, whose ideals still underlie the professional reward system in principle.

To be sure, the material rewards in agricultural research are not as great as in medicine and engineering, except for researchers working on "hot" technologies with big commercial potential. But it is the principle that is important, not just the amount of money. People who want to change the reward system in agriculture can ask its defenders whether their idea of the university is compatible with faculty members' engaging in remunerative outside activities linked to their academic positions. If they say no, shouldn't they give up such activities? If they say the two are compatible, how can they defend a reward system based on an ivory tower idea of the university that clearly is obsolete?

Individual Adaptation to the Dominant Reward System

Even within the traditional reward system, some people have been professionally successful while pursuing multidisciplinary research, long-term studies, or nontraditional research approaches. Perhaps the reward system already is flexible enough. Are major changes really needed?

The following are strategies for doing research that ordinarily would not be well rewarded:

1. First, become well established by traditional criteria; afterwards, work as you wish. This is an inadequate solution because it offers little to young researchers, the ones who often have the keenest interest in unconventional approaches. By the time researchers have enough seniority to avail themselves of this strategy, they may be too thoroughly socialized into prevalent styles to want to try something very different.

2. Find an institution that looks kindly on, or at least tolerates, a variety of research styles and that values various kinds of publications. Unfortunately, this choice can preclude a later move to an institution where narrower standards prevail. Again, this strategy presents a special problem for beginning researchers, who generally must work at several institutions for a few years at a time before settling down. Such people often cannot be choosy about what jobs they take.

3. Adjust the presentation of the work to make it consistent with the dominant reward system. A multidisciplinary team might report its research in several papers, each in a journal of a different discipline and each with a different lead author. However, this may obscure the integrated thinking underlying the project, and may be wholly unsuited for the kind of research we have called nondisciplinary, as opposed to multidisciplinary. Also, it contributes to the widely decried proliferation of "least publishable units," and the consequent bloating of the scientific literature (Culliton, 1988; Hamilton, 1990).

Clearly, these strategies are of limited use. At best, they create niches for alternative styles of work. But they accept the current system instead of questioning it as fundamentally as it deserves.

Alternative Grounds for Allocating Professional Rewards

When we think about alternatives to the current reward system, two questions arise. One is pragmatic: Can the system be made to fulfill its stated purpose better? The other is deeper: Is that purpose legitimate?

ELIMINATING IRRESPONSIBILITY

The first concern is easier to handle. At least we can do away with the nonsense that sometimes crops up when the ideal gets applied in practice. Pseudoprecise procedures, such as formulas for allocating credit among multiple authors, simply will not do. Counting papers instead of reading them is not compatible with the claim that research is a profession. Payment by the piece is for those who pick peas or stitch garments, and researchers' response to anyone who tries to defend it should be uncompromising. A poor tenure decision, for example, may either saddle the department with a dud for several decades or force out someone who would have been a valuable colleague. Such a decision should not be entrusted to people who are not willing to exercise their professional responsibilities.

Of course, methods of reward that can be applied mechanically are easier. No doubt that is the reason for "legal and accounting procedures which operate with criteria of 'performance' and 'productivity,' which are tailored for contract work and marketing but which are quite inapplicable, and indeed demonstrably detrimental, to scientific proficiency" (Weiss, 1964, p. 1209). Fortunately, a substantial reaction has been developing against the supremacy of quantity over quality (Mooney, 1991). On the other hand, earlier perceptions may linger despite this trend, causing researchers to feel pressured to meet an assumed publication quota even if it has lost some importance.

MATCHING THE REWARD SYSTEM TO THE GOALS OF RESEARCH

There is more to creating an appropriate reward system than doing away with mindless abuses. The task is not simply a matter of measuring accurately whether a researcher has met a particular goal; the goal itself can properly vary from one researcher to another. Even if the achievement is measured sensibly, a researcher's contribution to advancing disciplinary knowledge is relevant only in a field where the purpose is to advance disciplinary knowledge.

In agriculture, this often is not the goal. Most agricultural research is supported for other reasons, so advancement in disciplinary knowledge is not an appropriate measure of the work's value. Before there is further discussion of the reward system, therefore, the profession must agree on the right question: not "How well does this person's work meet a traditional standard for a researcher's 'output'?" but "How well has the work fulfilled the social goals for which it was supported?"

Moreover, "fulfillment of social goals" is an institution-wide matter, particularly for a public university. Today's land-grant university is very different from the traditional research university. It is not merely a home for independent subspecialties and autonomous researchers—it has a larger social mission. Fulfilling that mission requires a mix of departments with varying styles of work. If researchers' primary accountability is to their disciplinary colleagues, their work will not necessarily blend into a coherent whole that serves the university's overall purpose (Schuh, 1984).

ACCOMMODATING DIVERSITY

Matching professional rewards to the varied social goals of research means eliminating unjustified uniformity in the way a researcher's work is evaluated. Practices that were suitable in an earlier kind of university and that still are suitable in some other disciplines are not suitable for the full range of agricultural researchers.

This means, first, that the people asked to participate in judging a researcher's work need not be from the same field (Berk, 1981). People from other fields should be called upon to offer informed, professional opinions. Much agricultural research is not very arcane, and its value can readily be appreciated by researchers in other disciplines. If people in different disciplines find a piece of research very useful, their approval could be even more persuasive than an endorsement from a narrow circle in the researcher's own specialty.

Conversely, if the work investigates a topic that should interest researchers in other fields, we should be very concerned if their reaction is "So what?"

Also, we no longer should assume that valuable work must fit a particular mode. Whether its style is that of "basic" or "applied" science should not automatically argue for or against it. Nor should it automatically matter whether the research involves one discipline, several, or none at all; whether it was done at a working farm, experiment station, greenhouse, or laboratory; or whether the choice of topic reflected the values and preferences of farmers, researchers, or any other group. Whether the researcher worked alone or in collaboration should not matter either; the last person named on an eight-author paper may have more to be proud of than some sole authors.

Finally, we should not assume that the most worthwhile research gets published only in the few leading journals of the field. Being accepted for publication may involve more than quality: the paper also must conform to the editors' ideas of what constitutes worthwhile research. We have already noted some journals' preference for inferential statistics and for hypothesis-testing over exploratory or descriptive social science research. Also, researchers whose work is valuable for an especially broad range of fields—a characteristic that should be prized in agriculture—may want to publish it in a nondisciplinary journal with an appropriately broad readership. Unfortunately, even the best of such journals sometimes are disparaged as not truly professional outlets for serious research; the term "nondisciplinary" is sometimes (mistakenly) regarded as interchangeable with "non–peer-reviewed."

JUDGMENT AND OBJECTIVITY

In sum, when judging a researcher's contribution, our criteria should not be limited to a uniform ideal of professional accomplishments and a rigid procedure for measuring the fulfillment of that ideal. But if we accept this argument, what criteria are left?

> The answer may lie in giving human experience and judgment wider play. . . . There seems to be no better yardstick . . . than mature, sober, balanced judgment, taking into account the objectives of the insti-

tution, the special functions of its subdivisions . . . the variegated complexion of its faculty, its finite means and facilities. . . . [and] the principles of freedom for individual self-development and self-expression. (Weiss, 1964, p. 1214)

We endorse this principle—but with an important caution. "Mature, sober, balanced judgment" means exactly that. Saying that evaluation criteria can appropriately be judgmental is not equivalent to saying that anything goes. We must not condone arbitrariness, bias, or unfairness; otherwise the result could be worse than a system that uses objective criteria—like number of papers—that are less than fully appropriate but that at least are used fairly. Therefore, a reward system that is frankly based on judgment must include procedures that maximize its chances of getting the kind of judgment it wants.

In this way, it resembles a jury trial. A jury's verdict comes down in the end to the jurors' judgment. Nevertheless, jury trials have procedural safeguards: the right to challenge prospective jurors, restrictions on what may be introduced as evidence, the right of appeal, and a (presumably) unbiased person in charge. Analogous safeguards are appropriate for judging a professional's work.

Still, although no one disputes the importance of courtroom procedures, a jury's verdict remains a matter of judgment. Evaluations of researchers are even more so. In a courtroom trial, at least there *is* an objective answer; the only question is whether the jury's judgment comes up with it. In judging scholarship, the question itself—"How worthwhile is this person's work?"—is not objective. On one hand, the subjectivity of the question opens the door to abuse; on the other, it rules out purely "objective" standards as the only criteria.

The Prospects for Change

The approach we suggest for allocating professional rewards amounts to making the system flexible enough to handle the full range of legitimate research. How flexible should it be, and how strongly should the dominance of the traditional system be challenged?

The single term "agricultural research" encompasses widely varying activities, as we have stressed throughout. A system that governs a wide range of research by a narrow range of standards seems inappropriate.

This does not mean that a change in the system either can or should be carried out immediately. Considerable thought must be given to procedures that are both equitable and appropriate. The procedures will vary among disciplines and institutions; here we have tried only to inspire those who are discontent with the prevailing system to think about alternatives instead of simply resigning themselves to its supposedly inexorable constraints.

THE FORCES FOR CHANGE

Researchers sometimes blame others, such as administrators, for imposing an ill-suited system on them. However, the dominant reward system largely reflects researchers' own values. On one hand, this is discouraging; if the people affected by the system are happy with it, it is not likely to change drastically. On the other hand, a self-imposed system can be overturned more easily *if* the profession ever decides it wants to.

Although we believe that a substantial change is justified, it is difficult to envision that individual researchers could bring it about. This seems to say the situation is hopeless. But change need not be a matter for individuals. Professional societies constantly take up issues affecting their members' professional situation—for example, by providing forums where they can be debated and by reporting on institutions that are trying new approaches. Substantial changes already have occurred in several related matters, such as access to information used in tenure reviews and reduction of bias in hiring and promotions.

Furthermore, change does not have to be achieved across the entire profession at the same time. A department head or a dean can set a tone for a particular department or school. When people in other disciplines and institutions become aware of these changes, this recognition can help overcome their feeling that a particular reward system is inevitable.

An alternative that is introduced at a prestigious institution can be especially influential. It is encouraging, therefore, although it may seem paradoxical at first, that shifts from the traditional reward system are happening most often at influential first-rank universities where "pure" scholarship is strong (Mooney,

1991). This should make it even easier for others to make similar moves eventually, although so far that does not seem to be happening (Hamilton, 1990).

Another reason for thinking that change is possible is that agricultural research is increasingly susceptible to pressure from outside. Earmarked public research funds, as described in the next chapter, often impose high public accountability. Typically, outside groups are less than fully convinced by the arguments the profession uses to justify the prevailing reward system.

Private foundations are another significant funding source. Some are interested not only in supporting research on particular topics but also in influencing the arrangements under which research is done. For example, they may encourage greater involvement of outside groups and more outreach activities. Also, the research they support often does not have the goal of advancing disciplinary understanding. In times of hard-pressed budgets, universities cannot afford to ignore such funding. But they should be candid in acknowledging that their dependence reduces their professional autonomy. In other words, it undercuts a fundamental principle on which the reward system is based.

THE QUALITY QUESTION

We can expect some people to object to significant change on the grounds that it will jeopardize scientific quality (Beattie and Watts, 1987). For example, administrators sometimes are prejudiced against multidisciplinary research as shallow, whereas a disciplinary orientation supposedly guarantees quality (Saxberg et al., [1981] 1986).

Agricultural researchers working in alternative modes usually address this matter by reassuring potential critics that their work can be just as good. But they can go further by turning the critics' premise around: Are we sure that the prevailing system has guaranteed quality? If so, why did the criticisms in the Pound Report (National Resource Council, 1972) strike such a responsive note, and how much has agricultural research improved since then? Is it only ignorant prejudice that makes agricultural researchers such "poor cousins" among other scientists? Unfortunately, some people are immune to doubts about the prevailing quality of agricultural research, and they will object to reforming the system they believe has maintained that quality. But the issue is too important to be taken for granted.

The Reward System and Disciplinary Boundaries

Establishing an appropriate professional reward system could be made easier by the changes in departmental and disciplinary boundaries we suggested in the preceding chapter because research in a given department would become more uniform in style. Some departments would become more like basic science departments, or even merge with them. For these, the reward system could remain as it is.

In contrast, departments with a strong farm orientation have more reason to reform the system because their work is more likely to involve multidisciplinary and on-farm research and a greater role for farmers. Presumably, the members of such a department will share a high regard for such research and will adjust their professional standards accordingly. What will count is how well the research fulfills the department's goals, not whether it has a particular look that traditionally has been prized.

Of course, it is not enough for this attitude to prevail in only one department; the university, including the administration and the general faculty, must accept new ideas about professional rewards. But the reward system already has deviated from the traditional ideal, and further shifts are plausible as universities become even more dependent on outside political support and orient their programs accordingly. The more responsive they are to outside demands, the more diversified their programs will become, and the more they will need to adjust their reward system to handle this diversity.

Traditionally, such accommodation has been viewed with suspicion, especially where the values of the more inward-looking pure sciences prevail. The profession typically bemoans the need to adapt its governance to outside influences, claiming it will do a better job if left to take care of these purely professional matters itself. In the case of the reward system, adjusting to outside influences could benefit the profession itself, although the change might never have occurred if it had been left entirely to the profession.

12

. . .

Targeted Grant Programs

We were utterly ignored for a long time, then we were tolerated, and we are coming now to enjoy some of the symptoms of respectability. . . . But we are standing on dangerous ground. There will be a day of reckoning with us if we are not careful. . . . Before we put up much more for agricultural research is it not well to ask ourselves whether the work is being done in the best way?—E. Davenport, in Association of American Agricultural Colleges and Experiment Stations, 1907, p. 64

Advantages and Problems of Targeted Grant Programs

Programs that award grants in specific research areas have become increasingly popular at both the state and federal level and among private foundations. These programs complement two other important funding mechanisms: institutional funding, such as for the state experiment stations and the Agricultural Research Service; and broad grant programs in which researchers take the initiative in proposing specific projects within a general field, instead of responding to a detailed grantor-initiated solicitation.

The popularity of targeted grant programs is easy to understand. Agricultural research is under increasing external pressure to give more attention to topics such as reducing the use of environmentally damaging agrichemicals. A targeted program allows a quick and direct response because it does not imply a long-term financial or institutional commitment. Therefore it might face fewer bureaucratic obstacles than setting up a new department or institute.

These advantages may create other problems, however. It is easy to earmark funds for particular topics, but it also is easy to unearmark them—sometimes too easy. Flexibility is desirable, of course. But stopping a program when it no longer is needed is different from ending it because it no longer happens to be trendy.

When a program is funded out of a sense of urgency, people may demand that it produce results quickly to justify its favored status. Unfortunately, some research topics will not yield quick results: tracking environmental effects, resource depletion, and changes in soil productivity may require more than the few years typically allowed for a research project. Similarly, high visibility helps garner public support, but a highly visible program can become a political battleground where various groups try to promote partisan interests.

To avoid these problems while remaining timely and responsive, a program must give careful thought to several questions:
- The structure of the research teams and the style of research they pursue;
- The length of project funding;
- The suitability and qualifications of the researchers;
- Coordination among projects;
- External advice and review.

We discuss these points with particular reference to the research approaches discussed in part 2. Those approaches can be used in many kinds of research, but they will be especially suitable for many targeted grant programs. First, targeted research often should be multidisciplinary because the topics are determined by a "real world" problem, not by disciplinary considerations. Second, such programs may be especially eager to involve farmers and other "outside" groups in the research because many of the programs have been set up in response to pressures from such groups. Third, they often will favor research with an agroecological orientation because they give high priority to issues such as environmental protection and resource conservation.

Project Structure and Research Style

Besides setting forth goals and topics, program guidelines usually cover how the projects should be organized and managed. Sometimes the program fosters a particular way of working, as well as a particular topic. If so, the *how* and *who* of the program may be emphasized as much as the *what*.

For example, the guidelines may insist on a multidisciplinary research team, or a working farm as the research site, or a farmer advisory committee, and they might encourage collaborations between researchers and grass-roots organizations. Also, the program may promote more "holistic" or "systems" research to offset a perceived overemphasis on "reductionism" in other research. A program especially concerned with policy implications may expect the researchers to communicate their results to outside "user" groups.

Some researchers resist such requirements, perhaps because they do not want to be part of a multidisciplinary team or to collaborate with farmers. Outreach activities may not help them professionally and may even work against their professional stature in some circles. Also, researchers may be reluctant to publicize their work widely because they are afraid that advocacy groups will cite it inappropriately, ignoring its uncertainties and limitations.

Program managers must balance these preferences against their own ideas about research. If, instead, they specify exactly the kind of research they want, and researchers simply respond to it,

> the question is whether their response really matches the actual distribution of intellectual opportunities in science. . . . Without an accurate picture of the state of science, funding agencies run the risk of pouring public money into fields whose titles are politically appealing but which are not scientifically ripe for major advances. (Gustafson, 1975, p. 1065)

In requiring specific research approaches, there is a risk of overdoing it: style of research should not be elevated to the same status as the program's goals. Instead of requiring projects to be done on farmers' fields, for example, the guidelines might require the proposal to justify the choice of site, whether farmers' fields or otherwise. Guidelines that are too rigid can stifle imaginative research. It seems prudent to assume that they do not cover every worthwhile possibility, and that researchers might propose a style of project that serves the program's goals even better.

On the other hand, specific guidelines could validly be used to redress an imbalance or lack of diversity in existing research. Also, the flexibility we are calling for refers only to how the research is done, not to its goals. We have said repeatedly that public agricultural research is not "pure" science that should be allowed to set its own goals. Research priorities are a social and political matter, and program goals should reflect relevant public concerns (Bird, 1991, p. 36).

Besides leaving room for imaginative alternatives, flexible guidelines avoid another well-known problem: proposals that meet the letter, but not the spirit, of the rules. Researchers may see the new grant program simply as a source of money for work they want to do anyway. Thus they may repackage an old proposal so that it seems to meet the new requirements, perhaps by using the buzzwords they know the grantor is looking for. For example, a project might seem to fulfill a program's requirement of multi-disciplinarity if its participants come from the relevant disciplines. However, research that needs a more integrated mode requires carefully selected people who are committed to that approach and have the right attitudes to succeed in it.

The appropriate question in judging proposals, therefore, is not "Will the work be done in such-and-such a way?" but "Will the work be done in a way that is likely to serve the program's goals?" Of course, this makes the program managers' jobs more difficult than going down a checklist, so that good outside advice is critical.

Length of Funding

How long to fund projects is a critical decision for any grant program. Typically, commitments are made for no more than about three years at a time, and sometimes for only one year. This limit is prudent, especially when future budgets are uncertain. However, short-term funding may have a more political rationale, especially in a highly visible, publicly funded program: it means that the program's managers will have something to show to legislators who demand tangible results.

Unfortunately, agricultural research deals with natural processes that cannot be forced into a bureaucratically imposed schedule. Some investiga-

173

tions might never get done unless money is committed for more than a few years. For example, a study might need to run through several cycles of a multiyear rotation.

It is not enough to say that such a project can start with short-term funding and hope to get continuation funds later. Some studies will have little value unless they run to completion. Understandably, researchers would be reluctant to put several years into an effort that might not be permitted to bear fruit. Sometimes the preferences are reversed: the funding source wants to support long-term research, but the researchers shy away from it because the professional reward system pressures them to publish frequently. The effect of this pressure is seen in the proliferation of "preliminary" reports that are not followed by enough work to be definitive. (The problem is not unique to today's professional norms or practices, however. A prominent experiment station director early in this century made this comment: "Science . . . is not a 'report of progress' that shows little more than what the investigator hopes some time to prove and will take up again when he has the opportunity. . . . So obvious a truth would scarcely need stating, were it not for the fact that our scientific literature is submerged with increasing records of incomplete and inconclusive observations" [Jordan, 1908, pp. 131–32].)

Long-term funding clearly is riskier: if the project turns out to be ill-conceived, the program loses more than if a two- or three-year project fails. As a compromise, the program might make a commitment in principle to the full project but award funds for only one to three years at a time. Continued funding would depend on periodic review.

This approach differs from short-term funding in two ways. First, there is a presumption that continuation funds will be available unless something goes seriously wrong. The project should not have to compete afresh every few years, perhaps under new guidelines that no longer would give it high priority. Second, interim decisions on continuation funding should not consider whether the previous research was valuable for outside "users." The only consideration should be a reasonable likelihood of achieving worthwhile results.

Funding a many-year project, of course, rules out support for several other, shorter projects, a choice that should not be made casually. Especially in a topical program with a practical goal, it can be risky to exempt a project from the obligation to show tangible benefits quickly. Most researchers would love such license, but it should be bestowed carefully, where the work clearly needs many years and where the significance of the project justifies the risk.

Most important, the freedom to plan research over a longer period should be bestowed only on researchers who are likely to use it fruitfully. We agree with the call to "stand up for the principle of risk-taking by the conscientious, competent, disciplined, and dedicated investigator" (Weiss, 1964, p. 1210). But that call came with a significant warning:

> Can universities nowadays really in clear conscience testify to the fact that most of the research within their precincts is carried out with the fullest possible measure of conscience, deliberation, responsibility, and competence? . . . A certain laxity of aim and effort has given rise to much shoddy, inconsequential, redundant, uncritical, and ill-conceived research, the mainsprings of which may have been nothing more than that "soft money" was available to support it. (Weiss, 1964, p. 1210)

Qualifications of Researchers

Even for the general run of research proposals, not only long-term or speculative projects, the qualifications of the investigators are critical. Yet grantors do not always give this matter the care it deserves. Sometimes they rely too much on how the research team looks "on paper." Unfortunately, a list of publications at best provides only a partial indication of the authors' creativity or scientific discernment. This point, which should be obvious, is sometimes ignored.

The agricultural literature has its share of work to which the previous quotation about shoddy and inconsequential research applies. The fact of publication is no assurance that a paper met even minimum technical standards. Nor does it guarantee that the work was worth doing or, more generally, that it reflects the wisdom that makes good research more than just a craft. (This distinction is a theme in chapter 13, where we discuss the education of future researchers, arguing that education should not shortchange the development of students' scientific understanding and judgment in favor of purely technical skills.) The former head of Oak Ridge National Laboratory, hardly an antiscience Philistine, has put it strongly:

> As an arbiter of scientific taste and validity, scientific literature is deficient in two respects. First . . . nonsense is not so generally recognized. . . . Second, the scientific literature in a given field tends to form a closed universe; workers in a field, when they criticize each other, tend to adopt the same unstated assumptions. A referee of a scientific paper asks whether the paper conforms to the rules of the scientific community to which both referee and author belong, not whether the rules themselves are valid. (Weinberg, 1967, p. 70)

Grants may be awarded more according to what the applicant already has published than by intelligent judgment of the chances that the proposed work will be done well. If so, those who already have published the most are most likely to get grants that, in turn, will enable them to publish more, and so on. Generally, they deserve this opportunity, which means the system is working the way it should. But a mistake contributes to a cycle that may be difficult to break.

To help avoid this, principal investigators could be asked to submit one or two relevant papers, so that their qualifications can be evaluated by more than just the number and titles of their publications. The proposal itself should be reviewed not only as a statement of what the researchers plan to do but also as evidence of how well they are likely to do it. Some grant programs specifically make this point in their instructions to proposal reviewers.

This skeptical attitude toward proposers' qualifications has a positive side: it opens the system to prospective researchers who otherwise might have a hard time breaking in. These candidates include people in nonagricultural disciplines that border agricultural disciplines, such as sociology as opposed to rural sociology. Especially in exploratory or nondisciplinary work, strong grounding in a specific discipline is less important than scientific acumen, adaptability in learning new material, and a willingness and ability to communicate with people in other fields. An insightful proposal, whether or not its authors have the "right" résumés, is evidence that the individual or team is well qualified. This assumes, of course, that the program's managers and their advisers can distinguish real insight from slick grantsmanship.

Broadening the field of participants does more than let the program attract more good proposals. It also invests in the future by recruiting high-caliber investigators into agricultural research. A modest grant provides an entry into the field. If the researchers do that work well, they will face lower obstacles if they want to shift even further into agricultural research. Indeed, an important reason for starting federal competitive grants in 1977 was to expand the pool of talent from which agricultural research could draw (chap. 3).

Integration beyond the Single-Project Level

SELF-CONTAINED VERSUS CUMULATIVE RESEARCH

Topical programs offer a good opportunity to correct a common shortcoming in agricultural research: projects that are too self-contained, with too little relationship to other work. This problem to some extent can be solved by individual projects, but it sometimes can be handled best at the program level.

This kind of fragmentation is different from disciplinary specialization, where the problem is that the scope of individual investigations is inadequate. Instead, it is the failure to make linkages among different investigations of similar scope.

To be sure, research papers usually begin with some discussion of previous research. Nevertheless, after the methods, data, and conclusions have been presented, the discussion might never return to the literature that presumably motivated the current work. Research results often are presented as though they were self-contained. They are added to existing knowledge of the subject but not made a part of it.

For example, a project might study how tillage method and fertilizer rate affect the yield of some crop. Although this question has been studied countless times, the new research may only show the effects on yield in *this* climate, on *this* soil, and in *this* cropping system. The researchers may pay little attention to the differences between the new results and those obtained under other conditions; they may even neglect to report what these differences are.

REASONS FOR EXCESSIVE PARTICULARIZATION

It is understandable that many areas of agricultural research are highly particularized. Especially for field experiments, results will vary with local conditions. This greatly complicates the task of discovering patterns, making generalizations, or learning which variables are important.

A focus on local conditions is reinforced by the rhetoric of alternative agriculture. Some researchers, especially those with an agroecological orientation, stress how each farm—even each field—is unique. They emphasize that generalized information is not adequate for an individual farm: "We are ecologists and know better; we reject placeless farming and the troubles it brings. We know that provenance, soil types, ecotypes, landforms, and microclimates, all associated with place, are critical" (Ehrenfeld, 1987, p. 186).

But an alternative orientation should imply *greater* attention to integrating results obtained under varied conditions. Alternative farming systems are supposed to be based on agroecological principles, which can come only from cumulative, systematic knowledge. It is important to be site-specific when putting into practice a farming system based on agroecological principles, but the discovery of those principles depends on a broadly integrated analysis.

The point is not new:

> Every country, differing from other countries in its climate and temperature, in its soil, in its facility for procuring manures, in the character and supply of its labor, in its commercial and political relations, must be expected to have an agriculture in some respects peculiar to itself. . . . At the same time, the general principles of agricultural practice are every where the same. (Colman, 1856, p. 387)

This was a contentious issue for early experiment stations. The Office of Experiment Stations, a federal agency, often criticized the stations for being preoccupied with local conditions instead of integrating their work with that of other stations and other branches of science (Ferleger, 1990). Today, regional projects sometimes coordinate the work of several experiment stations (Barnes, 1982). This is not done as consistently as it deserves, however, perhaps because researchers overlook a distinction we have stressed: developing usable information about farming methods versus learning more about agricultural processes. Especially for the latter, the results of single projects rarely will be adequate by themselves.

THE BENEFITS AND COSTS OF MULTIPROJECT INTEGRATION

A good example of successful integration of many projects is the Universal Soil Loss Equation (USLE), a technique widely used in analyzing soil conservation strategies. This equation was developed from over 10,000 plot-years of measurements on rainfall erosion at forty-nine locations (Meyer, 1984; Wischmeier and Smith, 1978; Wischmeier, 1984). From this morass of unstandardized and sometimes inconsistent data emerged a procedure for predicting the rate of erosion under widely varying conditions, using a few basic variables such as rainfall, soil type, slope, and ground cover.

Besides giving more meaning to the results from which it was derived, the USLE also guides future researchers in deciding what to measure and how to report the results. This helps to integrate future work into the already considerable soil conservation literature. Of course, no equation that so drastically simplifies such a complex phenomenon can hope to be perfect; the USLE's chief developer, among others, has called attention to its limitations (Wischmeier, 1976). Ironically, the USLE's success might be its main shortcoming: when a general procedure is this convenient, it can be accepted so fully that researchers no longer think about whether it is appropriate for the intended application. Fortunately, researchers are continuing to reexamine and refine the USLE.

After-the-fact integration of the kind that led to the USLE is not always possible. Unless all the details are presented when the primary research is published, the original researcher may be the only person who can make detailed comparisons with other results beyond general statements of the form "These results are consistent with those of . . ." Also, it may not be possible to assimilate new results into existing knowledge unless the experiment has been designed with that purpose in mind; for example, the researcher may need to choose an experimental parameter specifically to be compatible with an earlier study. Otherwise, there would be no way to know whether different results reflected real variations in the phenomenon being studied or merely incidental differences in experimental conditions.

Although the benefits of integrating independent projects often will be great, the costs can be great also. Besides demanding researchers' time, it may require a compromise in the experimental plan. This is similar to the problem faced when researchers from different disciplines must compromise on a single site for a multidisciplinary study.

Multiproject coordination does not always entail such tradeoffs. Often, a project merely needs to take on some additional tasks, such as making more detailed data available. However, researchers may be afraid that if they make their data and methods available to others, they may be used inappropriately or without suitable acknowledgment.

FORMS OF MULTIPROJECT INTEGRATION

When its benefits outweigh the costs, integration can be fostered in various ways. They vary in how strongly the projects must interact, and whether the integration occurs during the research stage or afterwards.

Sometimes, after doing a project the way it would have been done regardless of integration, the researchers need only recognize that tying their results more closely to those of other researchers is worth the effort. At other times, as previously explained, the details of the research protocol—for example, the sampling scheme, data collection methods, or analytical techniques—must be adjusted to be compatible with previously published work. In still other cases, coordination is possible only among projects planned concurrently.

The last kind of integration is especially feasible in a targeted program.

First, the participants presumably are more homogeneous in their outlook and interests. They may be more willing to make the extra effort both because they see its value and because they already know and trust each other. Second, the program itself may arrange meetings to promote such coordination. Often the investigators welcome such activities, but the program can mandate them in any case. Formal coordination is less likely to be required under institutional funding or a general grant program, where the prevailing attitude is that such matters are better left to the researchers. In contrast, for programs with specific missions, the funders' attitude is more likely to be "Take it or leave it." Whether this *should* be their attitude is another matter. We have argued that the purpose of research should be to contribute to cumulative knowledge. However, there are various ways to achieve this. Our earlier discussion of why program guidelines should allow some latitude applies here, too.

Advice and Oversight

We have discussed several aspects of grant programs that require flexibility and judgment: project structure and research style, length of funding, qualifications of the researchers, and integration among projects. To help with such matters, program managers often solicit outside advice.

A familiar and inexpensive way to get such advice is to send individual proposals and draft reports to outside reviewers. A drawback is that reviewers deal only with individual projects, not the program as a whole.

A popular alternative is an advisory committee for the entire program. Such a committee might consist of technical experts similar to those who review individual proposals. Another common strategy is to identify the specific constituencies who have an interest in the program and to make sure they are all represented on the committee: farmers, consumers, environmentalists, and the like (Bird, 1991, p. 36). Sometimes programs allocate committee seats numerically, as by specifying that a certain number must be farmers. Unlike the intellectually "pure" peer review system, this approach acknowledges the political side of research funding.

The process can become too politicized, however. The committee should be regarded as a source of advice, not a governing board with decision-making authority. Nor should it be a device that lets program managers deflect criticism by pointing to the impeccably balanced committee that gave its approval.

Selecting committee members mainly by constituency also perpetuates the fiction that there is such a thing as *the* farmer perspective, or *the* consumer perspective, or *the* environmentalist perspective. At most, a farmer will offer good ideas of the kind that are more likely to occur to a farmer. Program managers should not deceive themselves into thinking that all farmers are represented by a single member—or by two or five or even ten farmer members. What the program managers do get if they have more than one well-chosen farmer member is more good ideas.

Of course, research is not untainted by politics, but an emphasis on how committee seats are distributed across interest groups invites a power struggle. The question "How can I help this program best serve my state [or region, or country]?" can degenerate too easily into "How can I help it best serve my organization?" Unfortunately, some groups consider that getting more research on their particular issue is automatically desirable because it shows that their cause has attained "agenda status." On the other hand, we can take encouragement from the experience of the advisory committees established under the Research and Marketing Act (RMA), described in chapter 3. USDA persuaded many committee members to consider the range of interests served by the program instead of just seeking funds for their own commodities (Mainzer, 1958).

To keep the committee from becoming self-serving, a statewide program could recruit members mainly from other states. These members must be carefully chosen to overcome the fear that, as outsiders, they will not know the local scene well enough to give sound advice. Also, the committee should be understood to be purely advisory. That way, it will not be perceived as outsiders taking over a program that ought to be controlled by people within the state.

One problem with "user" advisory committees is that users might not appreciate research that provides no immediate practical benefit, such as an improved production technique, but instead contributes toward a long-term benefit. This short-sightedness is true even among people who, by the very goals they are working toward, should be especially committed to a longer view:

> The strongest supporters of sustainable agriculture tend to . . . want research results and practical information tomorrow, if not sooner. There is, in other words, not much of a constituency within contemporary sustainable agriculture doctrine for basic, long-term agroecological research, upon which a more environmentally-benign agriculture 20 years hence will need to be built. (Buttel, 1992, p. 21)

Also, the importance of a practical benefit can change quickly because of external circumstances. For example, a sudden increase in the price of fertilizer will give greater urgency to research on alternatives to fertilizer. A research program should respond to changing circumstances, but it should not have to change as rapidly as the committee members' priorities. (Although this issue arose with the RMA committees, nonresearcher members became more supportive of long-range and basic research after serving on the committees [Mainzer, 1958].) A compromise between the peer review and advisory committee approaches is to have program-wide review and oversight, as in the latter, but to choose a more disinterested committee, as in the former. Ideally, these advisers will not be blind to the political aspects of the research. But they will not be so strongly involved that they cannot work toward the larger goals of the whole program. It is better not to allow narrowly partisan views on the committee than to hope to offset one form of self-interest with another.

Ultimately, the committee's value depends more on the caliber of its members than on their backgrounds or affiliations. The suggestions we have offered can help in making the right choice, but they do not insure it. Choosing a good group of advisers, and making the best use of their expertise, is a critical part of program management. Like every aspect of a good targeted program, it cannot be handled by prescription.

13
· · ·

Educating Future Agricultural Researchers

The great problem of the higher education now before us is how to integrate special-
ism with the totality of which it is a part. . . . We have made, not two, but five blades
of grass grow where but one grew before. Let us, without relaxing our aims and ef-
forts in this direction, set before ourselves the higher task of improving the humanity
of our people.—Buckham, 1907, pp. 45–46

In part 2, we discussed changes in how research is conducted. Here we consider how the next generation of researchers can be educated so that alternative approaches will be used to best advantage.

Several reforms in agricultural education are implicit in the earlier chapters. For example, some form of multidisciplinary education is appropriate if
researchers are to understand the perspectives of people in other disciplines.
But the new directions in research we have described are not the only issues
that agricultural institutions face. They are under pressure because of social
developments, financial stress, and shifts in the student body. These forces
can complement or impede the educational reforms we suggest. Therefore,
we set our discussion in this broader context.

We deal mainly with land-grant universities, which award most degrees
in agricultural science (96 percent of doctoral degrees in 1990 [National Research Council, 1991, p. 72].) We concentrate on graduate education, the
sieve through which most agricultural researchers pass. It is in graduate
school that future researchers learn what research questions and methods are

184

considered appropriate and, more generally, how to perform as agricultural professionals. Socialization happens both through the formal requirements of degree programs and through subtle and informal means. The professionals with whom graduate students work influence their values and interests and help form their commitment to particular research styles and topics.

Goals of Agricultural Education

Suggestions for reform must start with the main purpose of agricultural education: developing the student's creative and critical capacities. Simply covering specific subjects will not achieve this. Scientific education should foster the openness that allows scientists to question their own assumptions and give a fair hearing to others. Questioning the assumptions of a discipline requires the scientist to see beyond the sometimes arbitrary borders of the discipline. Agricultural education should provide both sound technical training in an agricultural discipline and a comprehension of its social context. It should be "defined in broadly liberal terms that embrace both scientific technology and social issues yet . . . still [be] directed to the individual" (Breimyer, 1986, pp. 67–68).

Discussions of goals for agricultural education today can draw from earlier experiences in humanities education. Writing after the particularly strong calls for reform raised during the 1960s, the Panel on Alternate Approaches to Graduate Education (1973, p. 14) cautioned that "the effort at opening up the university and the disciplines must be governed by a sense of proportion, and by attentive concern for certain necessary and fruitful discontinuities between life inside and outside institutions of learning." That is, the problems given to students should not just reflect today's "real-life" problems; a graduate education should train students to deal with concerns that are not foreseeable now (Association of Graduate Schools, 1976). Scientists need the knowledge and ability to adapt to rapid changes in agricultural research (Haney and Field, 1991) and to work under various conditions.

In setting educational priorities to enable them to do this, one should distinguish among wisdom, scientific understanding, and technical skills or crafts. The priority of graduate education should be to create a suitable environment for the blossoming of the student's wisdom—the acumen and insight that are the overriding goal of agricultural education. Therefore, although advanced education in agriculture involves considerable technical knowledge, it should be built on a strong liberal education that undergirds that knowledge. We suggest several ways to broaden the education of future agricultural researchers. Of course, they should not be expected to master all the fields relevant to their own discipline, but at least they should be knowledgeable and open enough to work with people in other disciplines and to understand their purposes and methods.

UNDERSTANDING SOCIAL CONCERNS

Social concerns are stressed in many recent statements about higher education in sustainable or alternative agriculture (Richards, 1988; Lamm, 1989). However, it is difficult to cover ecological processes, production systems, and social issues in a single program. Most often, social issues get short shrift (Lamm, 1989).

To understand the public's concerns about agriculture, students need some knowledge of the environmental and health effects of agricultural practices. They also need to know how agricultural policy is established and how economics and values enter that process.

Students also should appreciate the connections between U.S. and world agricultural systems and problems (Schuh, 1984) and the effects of U.S. agricultural policies on other countries. The U.S. economy is part of an international economy, and agricultural scientists operate in that context. This is a powerful argument for a good liberal education as the foundation of an advanced agricultural education: "In a world which is becoming increasingly interrelated and interdependent, knowledge of other languages, cultures and modes of thinking should be strongly encouraged" (Haney and Field, 1991, p. 162).

COMMUNICATION AND TEAMWORK ABILITIES

Another purpose of education is developing communication skills. Students should be able to explain technical matters in terms that can be understood by the public and by professionals in other disciplines. As agricultural scientists, they must be able to interact with representatives of outside groups concerned with agriculture, such as environmental, consumer, and producer organizations. This requires them to understand conflicting points of view, to speak and write clearly, and to negotiate without fueling antagonism.

The ability to work closely with professionals in other disciplines can overcome the problem of excessively specialized technical training (Richards, 1988). It also is crucial for understanding public concerns and the environmental aspects of agriculture.

UNDERSTANDING NATURAL SCIENCES

Besides more emphasis on social sciences and communication, agricultural students need a good grounding in natural sciences. The Pound Report, in its critical assessment of agricultural research, argued that

> to produce top flight agricultural scientists there should be little distinction between training in agriculture and training in the basic sciences. Agricultural research needs investigators with minds and training equal to those attracted to any other research area. (National Research Council, 1972, p. 58)

The divergence between education in agricultural science and basic sciences is aggravated by the insularity of the agricultural colleges and the prejudices against them, both at other universities and within their own. Overcoming this separation could induce good basic science students to work on agricultural questions and allow agricultural scientists to contribute more to basic sciences.

In recommending that agricultural students receive a more thorough basic science education, our purpose is only to deepen their understanding and to improve their ability to think scientifically when they become agricultural researchers. We are not suggesting that the agricultural sciences be made over into "basic" fields.

The Need for Balance in Education

Are the educational goals that we have described appropriate for all agricultural science students? For example, should all be expected to know something about the social issues affecting their disciplines, or is that essential only for those who plan to work in public policy or social sciences?

This question echoes debates about agricultural education in the early years of the land-grant system. A century ago, the influential dean of the Illinois College of Agriculture argued for a balance between sciences and humanities in undergraduate education:

> We need a new definition for an agricultural course. Let it be this: A broad and liberal education from the standpoint of rural life and its affairs. . . . Let it be in part intensely technical, but let it include also the sciences, history, economics, the humanities, if you will, but only let it be stimulating. (Davenport, 1897, p. 87)

A similar argument can be made today about graduate training.

Granted, it may seem foolish to tamper with a system that has trained outstanding agricultural scientists. Is it reasonable to require a brilliant young plant geneticist, for example, to learn about the economics and sociology of agriculture and about international agricultural issues, when this means taking time away from plant genetics?

We say yes, for several reasons, even though so unequivocal an answer might seem to contradict our repeated emphasis on flexibility and choice. Our earlier emphasis, however, concerned researchers; here we are talking about students who are not yet researchers. Although we are proposing a kind of uniformity for students, its purpose is to help them make decisions as future researchers. They need to be aware of the range of possibilities and to select intelligently among them. They also must be qualified to pursue whatever direction they eventually select.

As researchers, they will need a grounding in the social purposes of agricultural research to make responsible decisions about what research is important, because agricultural science is, and should be, an applied field. Those who are not attuned to social concerns are not likely to contribute effectively to the fundamental mission of land-grant universities: "to produce that knowledge needed to solve society's problems and to make for a better life for our citizens" (Schuh, 1984, p. 22).

In the future, it will be less and less common for researchers to work in isolation from other disciplines or from public pressures on research. To do their job, they will need to communicate effectively beyond their immediate circle of colleagues. Moreover, they will have roles besides researcher: many will serve on program review panels and advisory committees, and some will become deans and administrators. They will not be well prepared to carry out these broader functions if they are trained only in how to do research, without understanding its social and political sides.

As teachers, too, they will need to appreciate public concerns about agriculture because their students will raise such questions. They also will need some understanding of the international aspects of agriculture if they are to deal with the increasing proportion of students from other countries. The same is true for the many agricultural scientists who sometime in their careers will work for a company or agency involved in international affairs.

Admittedly, there are students for whom these educational goals will be irrelevant. However, as students they cannot be sure about their careers several decades into the future. Researchers' job preferences change, and changes sometimes are forced on them by unanticipated circumstances. Graduate school is too early to shut out career options.

Against this, one might argue that researchers can acquire a broader background later as the need arises. But it does not seem prudent to assume that researchers will devote themselves to new areas in mid-career. This point was made early in this century by the president of the Association of American Agricultural Colleges and Experiment Stations, in arguing why the land-grant colleges should require the study of languages, political science, economics, and literature, or, as he put it, "what one studies in order that he may know mankind at its best":

> What a wealth of thought and life he misses who is not introduced to its charms and potencies in the years when he is learning the possibilities of an intellectual life. If it be suggested that this lack can be made up by copious reading in after life, look around you and ask how many readers of great and good books you see among the half educated. (Buckham, 1907, pp. 43–44)

Also, students' careers will be influenced by what they are exposed to in school. A good graduate course on the social and environmental consequences of agricultural technologies might lead our hypothetical plant genetics student to get involved in setting policies for genetically altered organisms. It is circular to argue that students will not need to know anything about agricultural policy, and therefore offer them nothing that might awaken their interest:

> Much graduate education . . . intensif[ies] feelings of disengagement, of remoteness from community, and of chilling disbelief in the social uses of knowledge and imagination. The Ph.D. who is heedless of social reality and of the obligations and opportunities of specialized intelligence has, in a sense, been taught his heedlessness by the conventions of the system itself. (Panel on Alternate Approaches to Graduate Education, 1973, p. 30)

Social and Demographic Changes Affecting Agricultural Education

Anything done to advance the goals we have described must be compatible with the changing circumstances facing agricultural education. Of special concern are the number and origin of students, budget problems, and the influence of outside groups.

THE DECREASE IN THE TRADITIONAL SOURCES OF POTENTIAL STUDENTS

A significant change for agricultural colleges has been the problem of attracting enough good students. Historically, farm families have been the main source of agricultural students, but the number of people living on farms has declined steadily, both absolutely and as a fraction of the population (LeClere and Dahmann, 1990). The problem has been aggravated by the shift in the nonfarm population from rural to urban. Alternative sources of potential students are shrinking too: the proportion of the U.S. population between eighteen and twenty-four years old has dropped since 1980, and the number in this age range is not expected to return to the 1980 figure until the twenty-first century (Spencer, 1989, p. 7).

Enrollment problems are a sign of a greater challenge for agricultural education. Agriculture is playing a declining role in U.S. society, engaging only a small fraction of the workforce because of mechanization and increases in productivity. Current support for agricultural education is partly a legacy of agriculture's historical importance, not its present influence.

Agriculture is still a powerful force in many states, of course. Even so, federal and state funds have dropped in real terms (Campbell, 1991). These reductions have caused hiring cuts and have made universities reluctant to adopt reforms that are promising but untried.

EDUCATION FOR INTERNATIONAL AGRICULTURAL DEVELOPMENT

In contrast to shrinking employment opportunities in U.S. agricultural colleges, the demand for agricultural scientists qualified to work in developing countries is increasing because of rising population and greater development efforts. Almost two-fifths of recent recipients of U.S. agricultural science doctoral degrees are non-U.S. citizens, mainly from the developing world: from 1980 through 1990, the leading countries of origin (in descending order) were Taiwan, Brazil, Korea, Mexico, Nigeria, Iran, Thailand, India, and Iraq (National Science Foundation, 1991, p. 158). Most are here on temporary visas, although they will not all return home after graduation. U.S. students, too, are likely to be drawn increasingly to international work because of the changing job situation.

This trend may push agricultural science departments toward the needs of students preparing to work in developing rather than industrialized agriculture. Traditional agricultural science programs in the United States do not adequately prepare foreign students to work in their home countries (Swanson, 1986; Cashman and Persons, 1988). The problem goes beyond geographic and political differences that mean that students from other countries should know about topics such as management problems with tropical as opposed to temperate region soils, or subsistence farming systems as opposed to large-scale industrialized farming. There also are qualitative differences in how students from other countries should be educated. In some fields, the emphasis in the United States on disciplinary advances requires highly specialized studies and expensive equipment. However, the sponsors and employers of graduate students returning to developing countries often have different expectations. Typically, they want their scientists to have strong generalist backgrounds, to work with small budgets and simple equipment, and to contribute immediately to solving production problems (Swanson, 1986).

Some research reforms we have suggested are applicable not only in the United States but also to research for overseas development. For example, farmers' participation in research has been promoted especially strongly in agricultural development projects, where success often depends on farmers' commitment. On the other hand, training that gives more emphasis to the natural science foundations of agriculture might be seen as a luxury for a development program emphasizing immediate solutions of production problems. Such training can have long-term benefits, but it may be difficult to justify in a country that must find ways to feed more people immediately.

PRESSURES FROM INTEREST GROUPS

Another new pressure on agricultural educational institutions has come from environmental and food safety activists. These groups are among the strongest advocates of several ideas we have discussed, such as multidisciplinary research. They also are working for changes in education that reinforce the directions they want research to follow.

A related development is increased student interest in the environmental and health effects of agriculture (Hadwiger, 1982, pp. 48–49). Many students now enter graduate programs with some understanding of agriculture's environmental problems, and they want environmental science training to help them develop less harmful alternatives. Programs addressing environmental issues and agricultural sustainability are popular among students.

Along with the increasingly effective pressure of environmental and consumer groups, traditional interest groups still influence agricultural research through financial support and lobbying. However, the power balance among them has shifted. Commodity organizations are becoming more important than general farm organizations, which traditionally have been strong supporters of agricultural colleges.

Commodity organizations have little loyalty to their own state universities. Instead, they may contract research with any institution that can solve their particular problems (Campbell, 1991). Competition for commodity-oriented research funding affects the choice of thesis projects available to graduate students. It also may affect departments in a less tangible way, for example by influencing students' ideas about what questions are worthy of study. Heavy dependence on commodity-oriented research funds is likely to run counter to awareness of the social context of agriculture when students choose their research topics.

CHANGES IN THE STUDENT BODY

Demographic changes among agricultural students also have placed new demands on agricultural education. Nationality, race, urban versus rural background, and sex affect what students are interested in doing, what they become qualified to do, and what jobs are most likely to be open to them.

An important demographic change has been the increase in agricultural students with urban backgrounds. The proportion of agricultural scientists with farm backgrounds has been declining, although it still was 38 percent in 1979 (Busch and Lacy, 1983, p. 56). (This figure includes all scientists receiving USDA research support, whatever their field; it exceeds 50 percent if we limit it to strictly agricultural fields, such as agronomy, and exclude those that are less relevant here, such as forestry.)

Nonfarm students often want farm internships or practical on-farm experience. They may be sympathetic to farmers' points of view, but they probably will not identify with farmers in the same way as a person who grew up on a farm or whose relatives are still farming.

Another significant change is the larger proportion of women students: 16.2 percent of agricultural science doctoral recipients in 1980–90, compared with 1.5 percent in 1960–69 (National Science Foundation, 1991, pp. 23–24, 36). In their backgrounds and employment plans, female doctoral recipients are similar to their male counterparts (National Research Council,

1991, pp. 64, 66). However, social dynamics sometimes keep women and other nontraditional students out of educational and career opportunities, and attrition in science and engineering is higher for women than men, both from the bachelor's degree to the Ph.D. and from the Ph.D. to a job as a scientist or engineer (Office of Technology Assessment, 1986, pp. 131–39). No doubt this happens at least partly because women perceive themselves as being treated differently (Office of Technology Assessment, 1986, pp. 136–40). This problem is particularly pressing in agriculture because women have entered the field only recently and are present only in small numbers in most disciplines.

Diversifying the Researchers and the Institutions
THE BENEFITS OF A DIVERSE COMMUNITY
A theme of this book has been the importance of diverse modes of agricultural research. If researchers have varied backgrounds and interests, the profession will be more likely to challenge untested assumptions and arguments. A variety of backgrounds and viewpoints can help keep the profession vital and creative and assure that important areas are not neglected.

This argument is especially applicable to research concerned with the "culture of agriculture": the human side of the development and use of agricultural techniques. For example, in some developing countries, women and girls do most agricultural labor, but they are bypassed in research and extension efforts conducted by men and aimed at men. In industrialized economies, minority and women farmers and farm workers may get ignored by mainstream researchers.

Minority and women researchers are more likely to be attuned to the problems of minority and women farmers and farm workers. Moreover, in research undertaken cooperatively with farmers, establishing good rapport is easier if the researchers are from similar backgrounds, especially for research dealing with farmers' values and social relations. (This is not, however, an argument against the need to bring more women into U.S. agricultural research on the grounds that most U.S. farm operators are men. Women's role in agriculture is greatly understated by the statistics on farm operators, which count only one operator per farm even when other family members contribute significantly.)

Agricultural researchers remain demographically homogeneous despite the changes in the student body described earlier. For example, only 8 percent of all employed doctoral agricultural scientists in 1987 were nonwhite, and only 7 percent were women (National Science Foundation, 1988, p. 27). These proportions are lower than for scientists and engineers in general; the same is true for recent doctoral recipients (National Research Council, 1991, pp. 56–59). The homogeneity is even greater among higher-level agricultural scientists: fewer than 2 percent of full professors in agricultural sciences in 1987 were women (National Science Foundation, 1988, p. 42).

This homogeneity is aggravated by the agricultural sciences' isolation from other disciplines and other kinds of institutions (Mayer and Mayer, 1974; Hadwiger, 1982, pp. 43–47). Even though "the movement of knowledgeable individuals from one organization or type of institution to another is, perhaps, the most efficient way to transfer knowledge" (Busch and Lacy, 1983, p. 61), many agricultural scientists work in the university where they received at least part of their education, and most received two or more degrees from the same institution (Busch and Lacy, 1983, pp. 57–59).

ATTRACTING A DIVERSIFIED STUDENT BODY

Diversifying the research community of the future requires diversifying today's student body. As we have noted, agricultural colleges already have had to adjust to demographic change. We think they should go further. They should promote demographic diversity by encouraging nontraditional students to apply: students with urban backgrounds, women, and ethnic and national minorities. Second, where appropriate, they should adapt their programs to meet the needs of these students, both to attract them and to keep them. The quest for greater diversity should go beyond the requirements of nondiscrimination and affirmative action recruitment.

This would have additional advantages besides producing a more varied research community in the future. Nontraditional sources of students can help offset the decrease in the traditional pool of potential applicants. Also, the quality of agricultural research is likely to get better simply because the talent pool will be bigger. Quality has been a long-standing concern, and the problem no doubt will get worse if we continue to depend on a shrinking segment of the population to provide a large proportion of the student body. The educational changes we are suggesting make it even more important to attract high-caliber students. They will need to develop a greater breadth of understanding while continuing to receive specialized training, and they will be expected to learn some skills independently.

DIVERSITY: MORE THAN A MATTER OF RECRUITMENT

Recruiting people from nontraditional sources is only the beginning of an effective strategy for diversifying the student body. Success also requires an atmosphere that encourages such students to remain in the field despite cultural or demographic differences. This means, for example, dismantling the social forces that discourage women from beginning or continuing in scientific careers (Hall and Sandler, 1982, 1984; Office of Technology Assessment, 1986, pp. 136–40).

Some ways to do this are specific to women, such as an effective sexual harassment policy covering students and faculty, and highly visible administrative support to overcome stereotypical treatment of women. Women students often can benefit from programs that encourage them to seek leadership positions and that help them deal with subtle discrimination and the lower professional expectations that society places on them. Students from all nontraditional groups will benefit if the department or university offers extracurricular activities in which students of both sexes and from differing ethnic backgrounds are likely to participate.

Another part of an appropriate academic environment is a faculty whose diversity is suited to that of the student body. Even students from traditional groups will gain a breadth of perspective if their teachers come from varied backgrounds. Nontraditional students will especially benefit because nontraditional faculty members are better equipped to understand their perspectives and problems. Success in attracting and keeping students from nontraditional groups will make it easier to do so in the future, as more of these students take faculty positions.

We are not suggesting that minorities, women, or other students now represented in small numbers should be directed only to similar faculty members, for example, as their thesis committee members, academic advisers, or career counselors. Such channeling would perpetuate the barriers that have kept these students underrepresented, and would isolate nontraditional faculty members.

Rather, the goal of diversifying the faculty is to create an environment in which nontraditional students can become part of mainstream agricultural research. A diverse faculty helps set a tone of openness and acceptance. It shows students and faculty that the career aspirations of nontraditional students are legitimate and realistic, and that it is worth the effort to try to overcome the hurdles placed in their way by social stereotypes.

MEETING THE IMMEDIATE NEED FOR NONTRADITIONAL FACULTY MEMBERS

The increasing number of women students promises more women faculty members in the future (barring high attrition), but there is a need for more women on faculties now. The same need holds for other underrepresented groups; the problem may be temporary, but only if more students can be attracted from those groups, as has happened with women.

A way to meet the immediate need is for universities to establish collaborative teaching arrangements with women and members of other underrepresented groups who work in industry or government. Other students also will benefit from contact with professionals working in varied situations (Haney and Field, 1991).

DIVERSITY AMONG INSTITUTIONS

A complement to demographic diversity is education in a variety of intellectual environments: "Diversity of educational background can be an important stimulus for innovation and fresh perspectives" (Busch and Lacy, 1983, p. 61). Most agricultural scientists now are trained in land-grant agricultural colleges; agriculture should receive greater attention at other universities.

This recommendation does not imply a reduced role for land-grant colleges. Indeed, a distinguished early leader in the land-grant system, Liberty Hyde Bailey, made the same recommendation:

> The teaching of agriculture of college and university grade ought not to be confined to colleges of agriculture. All universities . . . will in time have departments of agriculture, if they are real universities, as much as they have departments of language or of engineering. They cannot neglect any fundamental branches of learning. (Bailey, 1909, p. 225)

Bailey's prediction has not been fulfilled, of course. But we are suggesting something more modest than developing a full agricultural curriculum. At first, some faculty members could turn their efforts to agriculturally related research topics. This would provide educational opportunities for graduate students, such as seminars, supervised independent study, and thesis research. Eventually, the school might offer some agriculturally oriented courses.

Non–land-grant institutions, both private and public, often are well qualified to deal with the generically scientific aspects of agriculture. Scientists in some nonagricultural disciplines would benefit from working on agricultural topics. This would enable them to tap the abundant body of research on economically important organisms and to do field studies under conditions that are more controlled than in natural settings. Scientists studying agricultural problems have made important advances in fields that are not exclusively agricultural, such as statistics, ecology, and microbiology. Unfortunately, agricultural scientists are separated from related disciplines by lower prestige and institutional isolation.

Scientists on both sides would benefit if the barriers were removed. Each kind of institution brings different strengths to the research effort. Agricultural colleges have well-established production-oriented research, and their extension services are a mechanism for incorporating farmers' viewpoints into research and education. Other institutions (non–land-grant universities and non-agricultural components of land-grant universities) may have strong natural and social science programs that can complement those of the agricultural colleges. Also, the various institutions operate under different external and institutional pressures that push each kind toward certain areas and away from others: "There may be a need, also, of a kind of agricultural work that can best be done in an institution that is independent of direct state support, and that is not at once responsible to popular will" (Bailey, 1909, p. 225).

Educational Activities

Meeting the goals of agricultural education requires balance among the components of the curriculum: production agriculture, other agricultural subjects, natural sciences, and social sciences and humanities. Similarly, broad-based courses must be balanced against the need for specialized knowledge.

Learning to communicate effectively and to understand the viewpoints of various groups comes from a suitable mix of classroom instruction and practical experience, such as internships or on-farm research. Learning what makes each mode of research appropriate can come from direct experience with different types of projects in varied settings.

We suggest various activities that can contribute to these goals. The best mix for an institution will depend on its size, the depth of its faculty, budgetary constraints, and the attitude of the administration.

CURRICULUM CHANGES

In the traditional sequence, students take courses outside their specialty as undergraduates or early in their graduate programs. That is, they get their biology, or economics, or whatever, "out of the way." Then they concentrate on their specialty, first in courses in their home department, then in their thesis research. The usual expectation is that the maturing professional will deal ever more exclusively with one discipline.

If the natural and social sciences are to make meaningful contributions to agricultural students' education, they cannot be dispensed with so easily. The goals we have put forth include understanding how one's specialty is related to real problems of agriculture. Thus a well-rounded education includes values, policy questions, environmental effects, and the human consequences of research. Instruction outside the student's specialty should stress interactions and connections among different areas of knowledge. It should not just take the indiscriminate "heap of stones" approach (Gusdorf, 1977, p. 588) that we have criticized as an inadequate way to handle multidisciplinarity.

To allow such connections to be treated in depth, some instruction in nonagricultural subjects should be concurrent with advanced study in an agricultural science. The introductory courses now offered in nonagricultural disciplines usually will not serve the purpose. Assigning an existing course in philosophy or public policy, say, could even be detrimental, because students and agricultural science faculty may perceive it as wasting time. An undergraduate "Introduction to Sociology" course would be too superficial for the needs of future agricultural professionals; it would not show how social forces affect agriculture, or how sociological methods might be used to study agricultural issues. Even an introductory graduate course in rural sociology might not be suitable if it is intended to be the first course in a complete rural sociology curriculum. Students from another field probably will not take additional rural sociology courses. They need a course that is self-contained, as opposed to introductory.

Subjects that are not customary in agricultural science curricula need to be covered through courses tailored for the purpose. For example, an institution might offer an anthropology course on comparative agricultural systems around the world. Such courses could be complemented by seminars on social issues affecting agriculture. A seminar on biotechnology might go beyond the technical question "What is possible?" to address ethical and social questions: What is desirable? Who is affected by various options? What are the tradeoffs between increased productivity and other consequences of the technology? Another worthwhile possibility is decision case studies that let students step into the shoes of a decision maker (Crookston, 1989), such as a dairy farmer who must decide how to deal with neighbors' complaints about manure.

These suggestions raise the question of who will teach such courses and supervise the seminars and case studies. Few instructors have sufficient training in both the technical and social sides of these subjects, but neither side should be treated superficially. One possibility is collaborative teaching. Another is bringing in outside professionals who deal with both the social and agricultural sides of various issues, who can serve as speakers and discussion leaders. Whatever the approach, it will be effective only if there are faculty members who are sincerely committed to it.

TRAINING IN MULTIDISCIPLINARY RESEARCH

The emphasis on multidisciplinary research is a strong force for change in graduate education. Discussions of agricultural education typically include the exhortation for greater exposure to "outside" disciplines. The curriculum changes we have suggested would give students such exposure. However, equipping students for multidisciplinary research beyond the "additive" mode (chap. 5) involves more than teaching them a broader range of subjects. They must learn to work with people who see the world differently and to integrate their contributions with those from other fields.

The solution is not as simple as taking a wider range of courses. Agricultural students already study other fields, either to get scientific background (for example, biology, chemistry, physics) or to learn techniques such as statistics and laboratory methods. If, despite this exposure, they are not prepared for multidisciplinary work, what should be changed?

A buzzword commonly heard in such discussions is "systems approach." Some version of this idea could help students make connections among interacting processes and different kinds of knowledge. To have lasting benefits, however, a systems approach requires more than just presenting students with the usual box-and-arrow or web diagrams showing how everything is connected to everything else.

Mere awareness of interconnections is not enough. Students first need to understand when a "systems approach" is possible and desirable. Sometimes the term is vacuous: too little is known to allow meaningful integration of a system's components. Where we understand the interactions among components, however, students can manipulate them to explore their consequences. Building a model of a specific system or working with an available model allows them to understand what components to include and the limitations of the data that describe them. Farm internships and practice in running a student farm serve a similar purpose; by becoming a participant in a system, the student learns from the inside how it works.

INTERNSHIPS

Internships that assign students to work with professionals can offer benefits similar to those of multidisciplinary interaction. Teamwork has become more common in agricultural research, especially outside academia. Students benefit from planning collaborative projects, carrying them out, and communicating their results. Internships are most valuable if students are supervised by faculty (Herring et al., 1990) and if they have realistic expectations of what such projects can and cannot provide (Panel on Alternate Approaches to Graduate Education, 1973). Internships should offer a choice of locations, such as a working farm or a nongovernmental organization. They also should offer a choice among modes of research: production and non-production topics, laboratory analysis and fieldwork, and studies done on experiment stations and on commercial farms.

Describing their internships in informal seminars helps students clarify what they have gotten from the work and gives them practice in communicating with people who have not had similar experiences. At the same time, other students will be learning from them. Other ways to improve communication skills include courses in writing and speaking to technical and nontechnical audiences, class presentations, required departmental seminars, and coaching before a student gives a conference presentation.

Internships or research projects on commercial farms offer advantages to agricultural science students beyond simply learning how to communicate and work with different people. For students without farm backgrounds, such projects are an opportunity to learn firsthand about the farm family's perspective and the constraints that farmers face. Working on farms will show students how farmers use information in making decisions, and how judgment, values, and other personal factors influence their adoption of new practices. Research projects involving several farms can show them the wide range of farmers' management abilities and perspectives. Interestingly, in an early example of formal agricultural education, the Storrs Agricultural School (now the University of Connecticut), farm work was required even for students with a farm background:

202

Nearly all our students thus far have been farmers' sons, measurably familiar with the common operations of the farm; but . . . all, by engaging in farm work, enjoy all the advantages to be gained by regular practice under competent supervision. (Armsby, 1883, pp. 23–24)

Is There Time for Everything?

Can all these dimensions of a comprehensive graduate agricultural education be addressed adequately without the program becoming either too long and unwieldy, or too superficial and lacking in rigor? Students cannot expand their programs to encompass everything we have said they need unless they give up something. Therefore, educators need to set clear priorities.

Earlier we stressed the importance of an environment in which the student's creativity and insight can develop, that is, one that nurtures wisdom. But for wisdom to be valuable, a professional scientist requires scientific knowledge. Acquiring and using scientific knowledge in turn requires crafts. Each aspect is important, but craft is subordinate to scientific knowledge, and scientific knowledge to wisdom. This point was appreciated even in the Storrs example of the training of farmers, for whom crafts presumably were even more important than for researchers:

Having, in the first year, laid a good foundation of general scientific knowledge, the student goes on in the second to build upon that foundation the superstructure of special training for his future work. . . . To study the applications of science to farming . . . we must first have some science to apply. (Armsby, 1883, pp. 22–24)

Students who have learned how to approach agricultural problems wisely and who have a base of scientific understanding can acquire most research skills they need, such as computer programming, laboratory methods, or laying out a field experiment. These discrete and specific skills largely involve rote learning. Morever, when they are needed, their lack is obvious. In contrast, learning about the social sciences and humanities—for which we rejected the "let them wait until they need it" approach—involves more understanding and judgment, and it implies a breadth of interest beyond the minimum needed to get by. Students are less likely to pursue these areas on their own, and comprehending such areas is less likely if there is no formal study first.

Spending less time on "how-to" training would allow the curriculum to stress the more difficult problem of learning why and when to use various research methods. It also would allow students to learn when a technique should *not* be used. With this understanding, students would be more likely to choose meaningful research problems and to try to solve them in an appropriate way. Otherwise, they may spend their careers searching for problems they can solve with whatever tool kits they already have. Such research is not likely to be very innovative: "No matter how searching in itself, a technical training is capable only of improving 'efficiency' within the already existing agricultural system" (Richards, 1988, p. 353).

Spending less time on crafts training in agricultural science curricula also allows an increased emphasis on the natural and social science foundations of those crafts. This approach would diminish the differences between the training of agricultural and nonagricultural scientists and help overcome a barrier to cooperation between the two.

Deemphasizing formal instruction in crafts does not imply a lowering of disciplinary rigor, even though some people equate advanced agricultural science education with greater command of disciplinary techniques. Rather, our recommendation would raise standards because it emphasizes the unique connotation of the Ph.D.: that its holder can do *original* research. This is what gives the Ph.D. its special status and reputation among advanced degrees; a professional degree, in contrast, testifies that its holder can provide certain services (Storr, 1973, chap. 6).

The ability to do original research comprises much more than a command of research techniques. It also implies creativity in proposing interesting and worthwhile questions, and the understanding needed to integrate one's results meaningfully into a body of cumulative knowledge. Conversely, many skills used in original research are not unique to Ph.D.-level researchers; computer programmers, laboratory technicians, and research farm managers often are better at such things. If the highest degree in agricultural education does not emphasize the scientific wisdom needed to do truly original research, perhaps we should call it something other than Ph.D. (Storr, 1973, chap. 6).

There is a place for instruction in the craft of research. However, its justification should be clearly understood: it complements and reinforces the rest of a graduate education. It does so by letting the abstractions "wisdom" and "scientific understanding" become concrete through application to specific problems. Also, good training equips students to continue acquiring crafts skills throughout their careers. Graduate education need not teach every technique that the student will ever need. It need not even prepare the student fully for an entry-level job. That is one purpose of a postdoctoral fellowship, which in principle is a learning opportunity, not a low-paid staff position.

Professionals can acquire techniques as they need them. Indeed, they must do so to avoid falling behind. For example, no one educated before the 1980s learned how to use microcomputers in school. Yet most research professionals do use them. Clearly, learning does not end in school. We are suggesting that the formal curriculum concentrate on the kind of learning that is best done that way. Training in crafts, in contrast, can be part of a research project, or can be covered through short courses or independent study. Such training would emphasize that acquisition of skills does not stop with graduate school but is part of continuing professional development.

How Might Institutions Bring about Change?

How educational programs get changed depends on the university's size, its goals, and the attitudes of its faculty and administration. For example, at a smaller institution, an innovative approach might be adopted across an entire department, or even an entire college; at a larger one, a mixture of innovative and traditional approaches makes more sense. An institution in a state where agriculture plays a minor role has an incentive to innovate more thoroughly. By offering something unique, it can attract more students from outside the state and more nontraditional students from within the state. Larger universities in major agricultural states usually have a firmer base of support and may thus feel less pressure to change.

If a university has a strong commitment to introducing new topics and teaching methods, three strategies are possible. The first is to allow change to percolate slowly through the program as the department hires people suited to the new approach. This requires well-defined goals so that new faculty members can be chosen accordingly. An advantage is that the department can change at whatever rate the budget allows. Also, it appeals to students who want a less traditional program without disrupting the entire department or infringing on professors' established territories.

The second option is to shift the approach of the entire College of Agriculture at once. Such a conversion obviously requires both strong administrative support and support within each department. Its feasibility increases if the college can hire several new faculty members to catalyze the effort. If successful, it lets students experience the curriculum as a cohesive program. However, a practical drawback is that individual professors are unlikely to change their courses and teaching styles significantly, despite a top-down directive.

The third option is to link changes in aims and teaching methods to specific subject areas. For example, internships and an emphasis on environmental and social aspects of agriculture might be features of a program in sustainable agriculture and ecology, or in agricultural science, technology, and resources, or in agricultural development. This approach allows students to interact with faculty members and students who are interested in the same subject, not just an alternative style of education. On the other hand, it risks isolating educational approaches that instead should permeate many areas of the curriculum.

An obstacle to creating such programs is the long-standing emphasis on specialization in graduate agricultural education. However, the approaches we are suggesting are not necessarily incompatible with specialization. As we explained in chapter 5, one kind of specialization results when complementary expertise from several departments is merged to create a new department or research center. Students who do research in such a program can get some benefits of well-planned multidisciplinarity despite the program's specialized focus. This possibility is greater in traditional fields that were formed by this kind of merger, as in the example we gave of dairy science, and in newer programs, such as sustainable agriculture, integrated pest management, and rural studies. However, faculty members and administrators may resent it when subjects previously assigned to their units are moved to a new program (Dillman, 1991). More important, if these programs are isolated within the university, they will fail to benefit from pertinent thinking in other disciplines.

The educational reforms we have suggested are not really new, a point we have underscored by quoting early agricultural educators. We have discussed the value of direct experience, of integrating rather than simply amassing information from different areas, and of preparing students to solve socially meaningful problems instead of just analyzing abstractions. These approaches all build on key ideas of a general humanistic education. If similar suggestions have been made so often, why do we repeat them?

To answer this question, we must go back to the changes in research that these suggestions are intended to support. The "new" research approaches are not new either, as shown by the quotations that begin every chapter. The question remains: Why have these approaches not been more widely adopted already, if they have been around for so long?

The prospects for constructive change are not as pessimistic as the question seems to imply. In some cases it is fitting that these approaches have not been adopted. They are not appropriate for every line of research; they should complement those already being used, not replace them.

However, even where appropriate, these approaches are not always used. They face institutional rigidity and restrictive professional standards that define proper research. Although such obstacles are a serious problem, they can be overcome. In Part 3, we have suggested ways to do so.

Even if the obstacles are removed, the new approaches will not flourish everywhere they are appropriate. Some researchers do not want to work in the ways we have suggested, and forcing them to do so would be futile.

That is why we end with the topic of agricultural education, the area of greatest hope. The value of future agricultural research will be determined above all by the people who do it: who they are, how they have been trained, and what they seek to achieve in their professional lives. With sound agricultural education today, tomorrow's researchers will be motivated and prepared to help U.S. agriculture achieve what the nation expects from it.

References

Allen, Patricia, Debra Van Dusen, Jackelyn Lundy, and Stephen Gliessman. 1991. "Integrating Social, Environmental, and Economic Issues in Sustainable Agriculture." *American Journal of Alternative Agriculture* 6(1): 34–39.

Altieri, Miguel A. 1983. *Agroecology: The Scientific Basis of Alternative Agriculture.* Berkeley: Division of Biological Control, Univ. of California.

———. 1989. "Agroecology: A New Research and Development Paradigm for World Agriculture." *Agriculture, Ecosystems and Environment* 27:37–46.

———. 1990. "Agroecology and Rural Development in Latin America." In *Agroecology and Small Farm Development,* edited by Miguel A. Altieri and Susanna B. Hecht, 113–18. Boca Raton, Fla.: CRC Press.

Altieri, Miguel A., David L. Glaser, and Linda L. Schmidt. 1990. "Diversification of Agroecosystems for Insect Pest Regulation: Experiments with Collards." In *Agroecology: Researching the Ecological Basis for Sustainable Agriculture,* edited by Stephen R. Gliessman, 70–82. New York: Springer-Verlag.

Anderson, Molly D. 1992. "Reasons for New Interest in On-Farm Research." *Biological Agriculture and Horticulture* 8(3): 235–50.

Anderson, Molly D., and William Lockeretz. 1991. *On-Farm Research Techniques: Report on a Workshop.* Greenbelt, Md.: Institute for Alternative Agriculture.

———. 1992. "Sustainable Agriculture Research in the Ideal and in the Field." *Journal of Soil and Water Conservation* 47(1): 100–104.

Armsby, H.P. 1883. "The Storrs Agricultural School—Course of Study." In *Sixteenth Annual Report of the Secretary of the Connecticut Board of Agriculture, 1882–83*, pp. 18–26. Hartford: State of Connecticut.

References

Arthur, J.C. 1895. "Development of Vegetable Physiology." *Science* 2:360–73.

Association of American Agricultural Colleges and Experiment Stations. 1907. "Scientific Investigation under Government Auspices." In *Proceedings of the Twentieth Annual Convention of the Association of American Agricultural Colleges and Experiment Stations,* 62–68. Bulletin 184. Office of Experiment Stations, U.S. Department of Agriculture. Washington, D.C.

Association of Graduate Schools. 1976. *The Research Doctorate in the United States.* Report to the Association of American Universities. Washington, D.C.

Atwater, W.O. 1878. "Farm Experiments with Fertilizers." In *Eleventh Annual Report of the Secretary of the Connecticut Board of Agriculture, 1877–88,* pp. 345–67. Hartford: State of Connecticut.

———. 1882. "Account of Field Experiments with Fertilizers." In *Fifteenth Annual Report of the Secretary of the Connecticut Board of Agriculture, 1881–82,* pp. 343–67. Hartford: State of Connecticut.

Azzi, Girolamo. 1956. *Agricultural Ecology.* London: Constable & Company.

Bailey, L.H. 1909. *The Training of Farmers.* New York: Century Co.

Baker, Brian P., and Douglas B. Smith. 1987. "Self Identified Research Needs of New York Organic Farmers." *American Journal of Alternative Agriculture* 2(3): 107–13.

Banfield, Edward E. 1949. "Planning Under the Research and Marketing Act of 1946; A Study in the Sociology of Knowledge." *Journal of Farm Economics* 31:48–75.

Barnes, John M. 1982. "Regional Coordination of Scientists' Initiatives in Interdisciplinary Research." In *Enabling Interdisciplinary Research: Perspectives from Agriculture, Forestry and Home Economics,* edited by Martha Garrett Russell, 135–38. Miscellaneous publication 19. Agricultural Experiment Station, Univ. of Minnesota, St. Paul.

Beattie, Bruce R., and Myles J. Watts. 1987. "The Proper Preeminent Role of Parent Disciplines and Learned Societies in Setting the Agenda at Land Grant Universities." *Western Journal of Agricultural Economics* 12(2): 95–103.

Bella, D.A., and K.J. Williamson. 1976. "Conflicts in Interdisciplinary Research." *Journal of Environmental Systems* 6(2): 105–24.

Ben-David, Joseph. 1971. *The Scientist's Role in Society: A Comparative Study.* Englewood Cliffs, N.J.: Prentice-Hall.

———. 1973. "How to Organize Research in the Social Sciences." *Daedalus* 102: 39–51.

Bennett, Hugh H. 1946. "A National Program of Soil Conservation." *Journal of Soil and Water Conservation* 1(1): 21–24, 29–34.

Berk, Richard A. 1981. "On the Compatibility of Applied and Basic Sociological Research: An Effort in Marriage Counseling." *American Sociologist* 16: 204–11.

Bird, Elizabeth Ann R. 1991. *Research for Sustainability? The National Research Initiative's Social Plan for Agriculture*. Walthill, Neb.: Center for Rural Affairs.

Blackwell, Gordon W. 1955. "Multidisciplinary Team Research." *Social Forces* 33(4): 367–74.

Boger, Robert P., and Virginia T. Boyd. 1982. "Institutional Policy and Operational Issues Affecting Interdisciplinary Research." In *Enabling Interdisciplinary Research: Perspectives from Agriculture, Forestry and Home Economics*, edited by Martha Garrett Russell, 87–93. Miscellaneous publication 19. Agricultural Experiment Station, Univ. of Minnesota, St. Paul.

Bohr, Niels. 1961. *Atomic Physics and Human Knowledge*. New York: Science Editions.

Bonnen, James T. 1983. "Historical Sources of U.S. Agricultural Productivity: Implications for R&D Policy and Social Science Research." *American Journal of Agricultural Economics* 65(5): 958–66.

———. 1986. "A Century of Science in Agriculture: Lessons for Science Policy." *American Journal of Agricultural Economics* 68(5): 1065–80.

Breimyer, Harold F. 1978. "Agriculture's Three Economies in a Changing Resource Environment." *American Journal of Agricultural Economics* 60:37–47.

———. 1986. "The Economic Returns of Agricultural Education." *Agricultural History* 60(2): 65–72.

Buckham, M.H. 1907. "Annual Address of the President of the Association." In *Proceedings of the Twentieth Annual Convention of the Association of American Agricultural Colleges and Experiment Stations*, 40–46. Bulletin 184. Office of Experiment Stations, U.S. Department of Agriculture. Washington, D.C.

Burkhardt, Jeffrey. 1991. "The Value Measure in Public Agricultural Research." In *Beyond the Large Farm: Ethics and Research Goals for Agriculture*, edited by Paul B. Thompson and Bill A. Stout, 79–105. Boulder, Colo.: Westview Press.

Busch, Lawrence, and William B. Lacy. 1983. *Science, Agriculture, and the Politics of Research*. Boulder, Colo.: Westview Press.

Buttel, Frederick H. 1992. "Environmentalization: Origins, Processes, and Implications for Rural Social Change." *Rural Sociology* 57(1): 1–27.

Buttel, Frederick H., and Lawrence Busch. 1988. "The Public Agricultural Research System at the Crossroads." *Agricultural History* 62(2): 303–24.

Buttel, Frederick H., and Michael E. Gertler. 1982. "Agricultural Structure, Agricultural Policy, and Environmental Quality: Some Observations on the Context of Agricultural Research in North America." *Agriculture and Environment* 7(2): 101–19.

Buttel, Frederick H., Gilbert W. Gillespie, Jr., Rhonda Janke, Brian Caldwell, and Marianne Sarrantonio. 1986. "Reduced-Input Agricultural Systems: Rationale and Prospects." *American Journal of Alternative Agriculture* 1(2): 58–64.

Cacek, Terry. 1984. "Organic Farming: The Other Conservation Farming System." *Journal of Soil and Water Conservation* 39(6): 357–60.

Callicott, J. Baird. 1988. "Agroecology in Context." *Journal of Agricultural Ethics* 1:3–9.

Campbell, Donald T. 1969. "Ethnocentrism of Disciplines and the Fish-Scale Model of Omniscience." In *Interdisciplinary Relationships in the Social Sciences,* edited by Muzafer Sherif and Carolyn W. Sherif, 328–48. Chicago: Aldine Publishing Co.

Campbell, Rex R. 1991. "The Land Grant Colleges of Agriculture." *Rural Sociologist* 11(2): 3–8.

Caplan, Arthur L., ed. 1978. *The Sociobiology Debate: Readings on Ethical and Scientific Issues.* New York: Harper & Row.

Carroll, C. Ronald, John H. Vandermeer, and Peter M. Rossett, eds. 1990. *Agroecology.* New York: McGraw-Hill Publishing Company.

Carson, Rachel. 1962. *Silent Spring.* Boston: Houghton Mifflin Company.

Cashman, Kristin, and Edgar Persons. 1988. "Improving the Relevance of Formal Education and Training in Preparing International Students As Change Agents for Low-Input Agriculture." *American Journal of Alternative Agriculture* 3(1): 35–42.

Catton, William R., Jr., and Riley E. Dunlap. 1980. "A New Ecological Paradigm for Post-Exuberant Sociology." *American Behavioral Scientist* 24(1): 15–47.

Chambers, Robert, Arnold Pacey, and Lori Ann Thrupp, eds. 1989. *Farmer First: Farmer Innovation and Agricultural Research.* New York: Bootstrap Press.

Clancy, Katherine L. 1986. "The Role of Sustainable Agriculture in Improving the Safety and Quality of the Food Supply." *American Journal of Alternative Agriculture* 1(1): 11–18.

Cobb, John B. 1984. "Theology, Perception, and Agriculture." In *Agricultural Sustainability in a Changing World Order,* edited by Gordon K. Douglass, 205–17. Boulder, Colo.: Westview Press.

Coleman, Eliot. 1985. "Toward a New McDonald's Farm." In *Sustainable Agriculture and Integrated Farming Systems,* edited by Thomas C. Edens, Cynthia Fridgen, and Susan L. Battenfield, 50–55. East Lansing: Michigan State Univ. Press.

Colman, Henry. 1856. *Agriculture and Rural Economy: From Personal Observation.* 5th ed. Vol. 1. Boston: Phillips, Sampson and Co.

Committee on Experiment Station Organization and Policy. 1907. "Report." In *Proceedings of the Twentieth Annual Convention of the Association of American Agricultural Colleges and Experiment Stations,* 74–78. Bulletin 184. Office of Experiment Stations, U.S. Department of Agriculture. Washington, D.C.

Conway, Gordon R. 1986. *Agroecosystem Analysis for Research and Development.* Bangkok: Winrock International Institute for Agricultural Development.

Cook, Kenneth A. 1985. *Agriculture and Conservation: What Prospects for a Merger into Regenerative Production Systems?* Ankeny, Iowa: Soil Conservation Society of America.

Cox, George W., and Michael D. Atkins. 1979. *Agricultural Ecology: An Analysis of World Food Production Systems.* San Francisco: W.H. Freeman and Company.

Craig, J. 1902. "Cooperation in Experimental Work Between the Station and the Farmer." In *Proceedings of the Fifteenth Annual Convention of the Association of American Agricultural Colleges and Experiment Stations,* 102. Bulletin 115. Office of Experiment Stations, U.S. Department of Agriculture. Washington, D.C.

Crews, Timothy E., Charles L. Mohler, and Alison G. Power. 1991. "Energetics and Ecosystem Integrity: The Defining Principles of Sustainable Agriculture." *American Journal of Alternative Agriculture* 6(3): 146–49.

Crookston, R. Kent. 1989. *A Case for Agriculture.* Department of Agronomy and Plant Genetics, Univ. of Minnesota, St. Paul.

Crosson, Pierre. 1989. "What Is Alternative Agriculture?" *American Journal of Alternative Agriculture* 4(1): 28–32.

Culliton, Barbara J. 1988. "Harvard Tackles the Rush to Publication." *Science* 241:525.

Danbom, David B. 1979. *The Resisted Revolution: Urban America and the Industrialization of Agriculture, 1900–1930.* Ames: Iowa State Univ. Press.

————. 1986a. "The Agricultural Experiment Station and Professionalization: Scientists' Goals for Agriculture." *Agricultural History* 60(2): 246–55.

————. 1986b. "Publicly Sponsored Agricultural Research in the United States from an Historical Perspective." In *New Directions for Agriculture and Agricultural Research: Neglected Dimensions and Emerging Alternatives,* edited by Kenneth A. Dahlberg, 107–31. Totowa, N.J.: Rowman & Allanheld.

————. 1992. "Research and Agriculture: Challenging the Public System." *American Journal of Alternative Agriculture* 7(3): 99–104.

Davenport, E. 1897. "The Exodus from the Farm." In *Proceedings of the Tenth Annual Convention of the Association of American Agricultural Colleges and Experiment Stations,* 82–87. Bulletin 41. Office of Experiment Stations, U.S. Department of Agriculture. Washington, D.C.

Dillman, Don. 1991. "Comments on Brown and Ranney." *Rural Sociologist* 11(2): 16–17.

Dobbs, Thomas L. 1987. "Toward More Effective Involvement of Agricultural Economists in Multidisciplinary Research and Extension Programs." *Western Journal of Agricultural Economics* 12(1): 8–16.

Dobbs, Thomas L., David L. Becker, and Donald C. Taylor. 1990. *Sustainable Agriculture Policy Analyses: South Dakota On-Farm Case Studies.* Economics staff paper no. 90–5. Economics Department, South Dakota State Univ., Brookings.

Dover, Michael J. and Lee M. Talbot. 1987. *To Feed the Earth: Agro-Ecology for Sustainable Development.* Washington, D.C.: World Resources Institute.

Dunlap, Riley E. 1980. "Paradigmatic Change in Social Science: From Human Exemptions to an Ecological Paradigm." *American Behavioral Scientist* 24(1): 5–14.

Dye, Franklin. 1899. "Agricultural Progress and Profit." In *Official Proceedings of the Nineteenth Annual Session of the Farmers' National Congress of the United States,* 46–55. Boston: Wright and Potter Printing Co.

Edwards, C.A. 1987. "The Concept of Integrated Systems in Lower Input/Sustainable Agriculture." *American Journal of Alternative Agriculture* 2(4): 148–52.

Edwards, C.A., R. Lal, P. Madden, R.H. Miller, and N.G. Creamer, eds. N.d. *International Conference on Sustainable Agricultural Systems, Columbus, Ohio, Sept. 19–23, 1988: Workshop Suggestions on Policies and Strategies.* Emmaus, Pa.: Rodale Institute.

Egerton, Frank N. 1976. "Ecological Studies and Observations Before 1900." In *Issues and Ideas in America,* edited by Benjamin J. Taylor and Thurmond J. White, 311–51. Norman: Univ. of Oklahoma Press.

Ehrenfeld, David. 1987. "Sustainable Agriculture and the Challenge of Place." *American Journal of Alternative Agriculture* 2(4): 184–87.

Eijsackers, H., and A. Quispel. 1988. "Ecology and Agronomy As Symbionts? An Introductory Overview." *Ecological Bulletins* (Copenhagen) 39:7–12.

Elkana, Yoseph O. 1991. "Participatory On-Farm Research: An International Perspective." In *Agricultural Research Institute Annual Meeting: Participatory On-Farm Research Concepts and Implications.* Agro-Ecology Program paper AE91–15. Univ. of Illinois, Urbana.

Ellen, Roy. 1982. *Environment, Subsistence and System: The Ecology of Small-Scale Social Formations.* Cambridge: Cambridge Univ. Press.

Elliott, Edward T., and C. Vernon Cole. 1989. "A Perspective on Agroecosystem Science." *Ecology* 70(6): 1597–1602.

Elster, Jon. 1983. *Explaining Technical Change: A Case Study in the Philosophy of Science.* Cambridge: Cambridge Univ. Press.

Exner, Rick. 1990. "Systems Research—How?" *Practical Farmer* 5(1): 13–15. Boone, Iowa: Practical Farmers of Iowa.

Experiment Station Committee on Organization and Policy. 1985. *Research Perspectives: Proceedings of the Symposium on the Research Agenda for the State Agricultural Experiment Stations.* Texas Agricultural Experiment Station, College Station.

Federal Energy Administration. 1976. *Energy and U.S. Agriculture: 1974 Data Base.* FEA/D-76/459. Office of Energy Conservation and Environment, and U.S. Department of Agriculture, Economic Research Service. Washington, D.C.

Ferleger, Lou. 1990. "Uplifting American Agriculture: Experiment Station Scientists and the Office of Experiment Stations in the Early Years After the Hatch Act." *Agricultural History* 64(2): 5–23.

Flach, Klaus W. 1990. "Low-Input Agriculture and Soil Conservation." *Journal of Soil and Water Conservation* 45(1): 42–44.

Francis, Charles 1990a. "Future Dimensions of Sustainable Agriculture." In *Sustainable Agriculture in Temperate Zones,* edited by Charles A. Francis, Cornelia Butler Flora, and Larry D. King, 439–66. New York: John Wiley & Sons.

———. 1990b. "Practical Applications of Low-Input Agriculture in the Midwest." *Journal of Soil and Water Conservation* 45(1): 65–67.

Francis, Charles, Richard R. Harwood, and James F. Parr. 1986. "The Potential for Regenerative Agriculture in the Developing World." *American Journal of Alternative Agriculture* 1(2): 65–74.

Francis, Charles, and James King. 1988. "Cropping Systems Based on Farm-Derived, Renewable Resources." *Agricultural Systems* 27: 67–75.

Francis, Charles, James King, Jerry DeWitt, James Bushnell, and Leo Lucas. 1990. "Participatory Strategies for Information Exchange." *American Journal of Alternative Agriculture* 5(4): 153–60.

Gardner, John C. 1990. "Responding to Farmers' Needs: An Evolving Land Grant Perspective." *American Journal of Alternative Agriculture* 5(4): 170–73.

Geiger, Roger L. 1986. *To Advance Knowledge: The Growth of American Research Universities, 1900–1940.* New York: Oxford Univ. Press.

Gerber, John M. 1989. "Agroecology Defined." *Illinois Research* 31(3/4): 36.

———. 1990. "Agro-ecology Science Fuses Agriculture, Ecology." *Agro-ecology* 2(2): 1. Univ. of Illinois, Urbana.

Gillespie, Gilbert W., Jr., and Frederick H. Buttel. 1989. "Farmer Ambivalence toward Agricultural Research: An Empirical Assessment." *Rural Sociology* 54(3): 382–408.

Gliessman, Stephen R. 1984. "An Agroecological Approach to Sustainable Agriculture." In *Meeting the Expectations of the Land: Essays in Sustainable Agriculture and Stewardship,* edited by Wes Jackson, Wendell Berry, and Bruce Colman, 160–71. San Francisco: North Point Press.

———. 1987. "Species Interactions and Community Ecology in Low External-Input Agriculture." *American Journal of Alternative Agriculture* 2(4): 160–65.

Gliessman, Stephen R., ed. 1990a. *Agroecology. Researching the Ecological Basis for Sustainable Agriculture.* New York: Springer-Verlag.

Gliessman, Stephen R. 1990b. "Quantifying the Agroecological Component of Sustainable Agriculture: A Goal." In Gliessman, ed. *Agroecology: Researching the Ecological Basis for Sustainable Agriculture,* 366–70.

Goodell, Henry H. 1899. "The Mission of the Experiment Station." In *Official Proceedings of the Nineteenth Annual Session of the Farmers' National Congress of the United States,* 20–27. Boston: Wright and Potter Printing Co.

Granatstein, D. 1988. *Reshaping the Bottom Line: On-Farm Strategies for a Sustainable Agriculture.* Lewiston, Minn.: Land Stewardship Project.

Gusdorf, Georges. 1977. "Past, Present and Future in Interdisciplinary Research." *International Social Science Journal* 29(4): 580–600.

Gustafson, Thane. 1975. "The Controversy over Peer Review." *Science* 190:1060–66.

Hadwiger, Don F. 1982. *The Politics of Agricultural Research.* Lincoln: Univ. of Nebraska Press.

———. 1984. "US Agricultural Research Politics: Utopians, Utilitarians, Copians." *Food Policy* 9(3): 193–205.

Hall, Roberta M., and Bernice R. Sandler. 1982. *The Classroom Climate: A Chilly One for Women?* Project on the Status and Education of Women. Washington, D.C.: Association of American Colleges.

———. 1984. *Out of the Classroom: A Chilly Campus Climate for Women?* Project on the Status and Education of Women. Washington, D.C.: Association of American Colleges.

Hamilton, David P. 1990. "Publishing by—and for?—the Numbers." *Science* 250:1331–32.

Haney, Wava G., and Donald R. Field. 1991. "Charting a Course for 2020." In *Agriculture and Natural Resources: Planning for Educational Priorities for the Twenty-first Century,* edited by Wava G. Haney and Donald R. Field, 157–79. Boulder, Colo.: Westview Press.

Hardesty, Donald L. 1980. "The Ecological Perspective in Anthropology." *American Behavioral Scientist* 24(1): 107–24.

Hardin, Charles M. 1955. *Freedom in Agricultural Education.* Chicago: Univ. of Chicago Press.

Harding, T. Swann. 1940. "Science and Agricultural Policy." In *Farmers in a Changing World: The Yearbook of Agriculture, 1940,* pp. 1081–1110. U.S. Department of Agriculture. Washington, D.C.

Harwood, Richard R. 1984. "Organic Farming Research at the Rodale Research Center." In *Organic Farming: Current Technology and Its Role in a Sustainable Agriculture,* edited by D.F. Bezdicek and J.F. Power, 1–17. Madison, Wis.: American Society of Agronomy, Crop Science Society of America, and Soil Science Society of America.

———. 1985. "The Integration Efficiencies of Cropping Systems." In *Sustainable Agriculture and Integrated Farming Systems,* edited by Thomas C. Edens, Cynthia Fridgen, and Susan L. Battenfield, 64–75. East Lansing: Michigan State Univ. Press.

Hendrix, Paul F. 1987. "Strategies for Research and Management in Reduced-Input Agroecosystems." *American Journal of Alternative Agriculture* 2(4): 166–72.

Herring, Matthew D., Clark J. Gantzer, and Gregory A. Nolting. 1990. "Academic

Value of Internships in Agronomy: A Survey." *Journal of Agronomy Education* 19:18–20.

Heyer, Paul. 1982. *Nature, Human Nature, and Society: Marx, Darwin, Biology, and the Human Sciences.* Westport, Conn.: Greenwood Press.

Hightower, Jim. [1973] 1978. *Hard Tomatoes, Hard Times.* Reprint. Cambridge, Mass.: Schenkman Publishing Co.

Hildebrand, Peter E., and Federico Poey. 1985. *On-Farm Agronomic Trials in Farming Systems Research and Extension.* Boulder, Colo.: Lynne Rienner Publishers.

Hodges, R.D. 1981. "An Agriculture for the Future." In *Biological Husbandry: A Scientific Approach to Organic Farming,* edited by B. Stonehouse, 1–14. London: Butterworths.

Hopkins, C.G. 1904. "The Present Status of Soil Investigation." In *Proceedings of the Seventeenth Annual Convention of the Association of American Agricultural Colleges and Experiment Stations,* 95–104. Bulletin 142. Office of Experiment Stations, U.S. Department of Agriculture. Washington, D.C.

Howard, Sir Albert. [1943] 1976. *An Agricultural Testament.* New York: Oxford Univ. Press. Reprint. Emmaus, Pennsylvania: Rodale Press.

Huang, H.T. 1988. "The USDA's Competitive Grants Program and Agricultural Research." *Agricultural History* 62(2): 270–78.

Hurlbert, Stuart H. 1984. "Pseudoreplication and the Design of Ecological Field Experiments." *Ecological Monographs* 54(2): 187–211.

Jackson, Wes. 1990. "Agriculture with Nature As Analogy." In *Sustainable Agriculture in Temperate Zones,* edited by Charles A. Francis, Cornelia Butler Flora, and Larry D. King, 381–422. New York: John Wiley & Sons.

Jackson, Wes, and Jon Piper. 1989. "The Necessary Marriage Between Ecology and Agriculture." *Ecology* 70(6): 1591–93.

Janke, Rhonda, and Ken McNamara. 1988. "Using Replicated On-Farm Research Trials to Answer Farmers' Questions About Low-Input Cropping Systems." *Proceedings of Farming Systems Research/Extension Symposium: Contributions of FSR/E Towards Sustainable Agricultural Systems.* Farming Systems Research Paper Series no. 17. Univ. of Arkansas and Winrock International Institute for Agricultural Development.

Jaschik, Scott. 1991. "Political Activists Work to Change Land-Grant Colleges." *Chronicle of Higher Education* 37 (March 20): A1, A24.

Jenkins, Merle T. 1936. "Corn Improvement." In *Yearbook of Agriculture, 1936,* pp. 455–522. U.S. Department of Agriculture. Washington, D.C.

Johnson, Glenn L. 1971. "The Quest for Relevance in Agricultural Economics." *American Journal of Agricultural Economics* 53(5): 728–39.

———. 1984. *Academia Needs a New Covenant for Serving Agriculture*. Special Publication. Mississippi Agricultural and Forestry Experiment Station, Mississippi State Univ.

Johnson, S.W. 1882. Comments on paper by E.L. Sturtevant. In *Fifteenth Annual Report of the Secretary of the Connecticut Board of Agriculture, 1881–82*, pp. 55–59. Hartford: State of Connecticut.

Jordan, John Patrick, Paul F. O'Connell, and Roland R. Robinson. 1986. "Historical Evolution of the State Agricultural Experiment Station System." In *New Directions for Agriculture and Agricultural Research: Neglected Dimensions and Emerging Alternatives*, edited by Kenneth A. Dahlberg, 147–62. Totowa, N.J.: Rowman & Allanheld.

Jordan, Whitman H. 1908. "The Authority of Science." In *Semi-Centennial Celebration of Michigan State Agricultural College*, 128–45. East Lansing: Michigan State Agricultural College.

Kaplan, Norman. 1964. "Sociology of Science." In *Handbook of Modern Sociology*, edited by Robert E. L. Faris, 852–81. Chicago: Rand McNally & Co.

Kerr, Norwood Allen. 1988. "Institutionalizing the New Agenda: The State Agricultural Experiment Stations, 1977–1981." *Agricultural History* 62(2): 279–95.

Kirkendall, Richard S. 1966. *Social Scientists and Farm Politics in the Age of Roosevelt*. Columbia: Univ. of Missouri Press.

———. 1986. "The Agricultural Colleges: Between Tradition and Modernization." *Agricultural History* 60(2): 3–21.

Kirschenmann, Frederick. 1991. "Fundamental Fallacies of Building Agricultural Sustainability." *Journal of Soil and Water Conservation* 46(3): 165–68.

Kloppenburg, Jack, Jr. 1991. "Social Theory and the De/Reconstruction of Agricultural Science: Local Knowledge for an Alternative Agriculture." *Rural Sociology* 56(4): 519–48.

Koepf, H.H. 1981. "The Principles and Practices of Biodynamic Agriculture." In *Biological Husbandry: A Scientific Approach to Organic Farming*, edited by B. Stonehouse, 237–50. London: Butterworths.

Krome, Margaret. 1988a. *The Southwest Wisconsin Farmers' Research Network—1986–1987: A Two-Year Case History of an On-Farm Research Project*. Black Earth: Wisconsin Rural Development Center.

―――. 1988b. "Sustainable Research Standards Can Guide C A L S." *Newsletter* 5(3): 3. Black Earth: Wisconsin Rural Development Center.

Lamm, Tom. 1989. *Guidelines for Developing University Sustainable Agriculture Programs: Suggestions for Research, Extension and Instruction.* Black Earth: Wisconsin Rural Development Center.

LeClere, Felicia, and Donald Dahmann. 1990. *Residents of Farms and Rural Areas: 1989.* Current Population Reports, Population Characteristics. Series P-20, no. 446. Economic Research Service, U.S. Department of Agriculture, and Bureau of the Census, U.S. Department of Commerce. Washington, D.C.: U.S. Government Printing Office.

Lengnick, Laura L., and Larry D. King. 1986. "Comparison of the Phosphorus Status of Soils Managed Organically and Conventionally." *American Journal of Alternative Agriculture* 1(3): 108–14.

Levins, Richard. 1973. "Fundamental and Applied Research in Agriculture." *Science* 181: 523–24.

Levins, Richard, and John H. Vandermeer. 1990. "The Agroecosystem Embedded in a Complex Ecological Community." In *Agroecology,* edited by C. Ronald Carroll, John H. Vandermeer, and Peter M. Rossett, 341–62. New York: McGraw-Hill Publishing Company.

Lewin, Ralph A., and Nicholas Polunin. 1990. "Ecology and 'Ecostasis.' " *Environmental Conservation* 17(2): 177.

Lockeretz, William. 1985. "U.S. Organic Farming: What We Can and Cannot Learn from On-Farm Research." In *Sustainable Agriculture and Integrated Farming Systems,* edited by Thomas C. Edens, Cynthia Fridgen, and Susan L. Battenfield, 96–104. East Lansing: Michigan State Univ. Press.

―――. 1987. "Establishing the Proper Role for OnFarm Research." *American Journal of Alternative Agriculture* 2(3): 132–36.

―――. 1988. "Open Questions in Sustainable Agriculture." *American Journal of Alternative Agriculture* 3(4): 174–81.

―――. 1990. "What Have We Learned About Who Conserves Soil?" *Journal of Soil and Water Conservation* 45(5): 517–23.

―――. 1991a. "Information Requirements of Reduced-Chemical Production Methods." *American Journal of Alternative Agriculture* 6(2): 97–103.

―――. 1991b. "Multidisciplinary Research and Sustainable Agriculture." *Biological Agriculture and Horticulture* 8(2): 101–22.

————. 1991c. "The Organization and Coverage of Research on Reduced Use of Agricultural Chemicals." *Agriculture, Ecosystems and Environment* 36: 217–34.

Lockeretz, William, and Molly D. Anderson. 1990. "Farmers' Role in Sustainable Agriculture Research." *American Journal of Alternative Agriculture* 5(4): 178–82.

Lockeretz, William, and Patrick Madden. 1987. "Midwestern Organic Farming: A Ten-Year Follow-up." *American Journal of Alternative Agriculture* 2(2): 57–63.

Loucks, Orie L. 1977. "Emergence of Research on Agro-Ecosystems." *Annual Review of Ecology & Systematics* 8:173–92.

Lowrance, Richard, Paul F. Hendrix, and Eugene P. Odum. 1986. "A Hierarchical Approach to Sustainable Agriculture." *American Journal of Alternative Agriculture* 1(4): 169–73.

Lowrance, Richard, Benjamin Stinner, and Garfield House, eds. 1984. *Agricultural Systems: Unifying Concepts.* New York: John Wiley & Sons.

MacRae, Rod J., Stuart B. Hill, John Henning, and Guy R. Mehuys. 1989. "Agricultural Science and Sustainable Agriculture: A Review of the Existing Scientific Barriers to Sustainable Food Production and Potential Solutions." *Biological Agriculture and Horticulture* 6(3): 173–219.

MacRae, Rod J., Stuart B. Hill, Guy R. Mehuys, and John Henning. 1990. "Farm-Scale Agronomic and Economic Conversion from Conventional to Sustainable Agriculture." *Advances in Agronomy* 43: 155–98.

Madden, J. Patrick. 1986. "Toward a New Covenant for Agricultural Academe." In *The Agricultural Scientific Enterprise: A System in Transition,* edited by Lawrence Busch and William B. Lacy, 267–79. Boulder, Colo.: Westview Press.

————. 1989. "What Is Alternative Agriculture?" *American Journal of Alternative Agriculture* 4(1): 32–34.

Madden, J. Patrick, and Paul O'Connell. 1989. "Early Results of the LISA Program." *Agricultural Libraries Information Notes.* Vol. 15, no. 6/7. National Agricultural Library, U.S. Department of Agriculture, Beltsville, Md.

————. 1990. "LISA: Some Early Results." *Journal of Soil and Water Conservation* 45(1): 61–64.

Magdoff, Fred. 1991. "Managing Nitrogen for Sustainable Corn Systems: Problems and Possibilities." *American Journal of Alternative Agriculture* 6(1): 3–8.

Mainzer, Lewis C. 1958. "Science Democratized: Advisory Committees on Research." *Public Administration Review* 18: 314–23.

Mangelsdorf, Paul C. 1951. "Hybrid Corn." *Scientific American* 185(2): 39–47.

Marcus, Alan I. 1986. "From State Chemistry to State Science: The Transformation of the Idea of the Agricultural Experiment Station, 1875–1887." In *The Agricultural Scientific Enterprise: A System in Transition,* edited by Lawrence Busch and William B. Lacy, 3–12. Boulder, Colo.: Westview Press.

Marshall, Eliot. 1990a. "Data Sharing: A Declining Ethic?" *Science* 248: 952, 954–55, 957.

———. 1990b. "When Commerce and Academe Collide." *Science* 248: 152–56.

Mayer, André, and Jean Mayer. 1974. "Agriculture, the Island Empire." *Daedalus* 103(3): 83–95.

McCalla, Alex F. 1978. "The Politics of the U.S. Agricultural Research Establishment." *Policy Studies Journal* 6(4): 479–83.

McCollum, E.V., Elsa Orent-Keiles, and Harry G. Day. 1939. *The Newer Knowledge of Nutrition.* 5th ed. New York: Macmillan Company.

McIntosh, Robert P. 1976. "Ecology Since 1900." In *Issues and Ideas in America,* edited by Benjamin J. Taylor and Thurmond J. White, 353–72. Norman: Univ. of Oklahoma Press.

———. 1985. *The Background of Ecology: Concept and Theory.* Cambridge: Cambridge Univ. Press.

———. 1989. "Citation Classics of Ecology." *Quarterly Review of Biology* 64(1): 31–45.

Meine, Curt. 1987. "The Farmer As Conservationist: Aldo Leopold on Agriculture." *Journal of Soil and Water Conservation* 42(3): 144–49.

Meyer, L.D. 1984. "Evolution of the Universal Soil Loss Equation." *Journal of Soil and Water Conservation* 39(2): 99–104.

Molnar, Joseph J., Patricia A. Duffy, Keith A. Cummins, and Edzard Van Santen. 1992. "Agricultural Science and Agricultural Counterculture: Paradigms in Search of a Future." *Rural Sociology* 57(1): 83–91.

Montagu, Ashley, ed. 1980. *Sociobiology Examined.* New York: Oxford Univ. Press.

Mooney, Carolyn J. 1991. "Efforts to Cut Amount of 'Trivial' Scholarship Win New Backing from Many Academics." *Chronicle of Higher Education* 37(May 22): A1, A13, A16.

National Research and Education Program on Sustainable Agriculture. 1991. *Guidelines for Grants Programs on Sustainable Agriculture.* National Research and Education Program on Sustainable Agriculture, Northeast Region.

National Research Council. 1972. *Report of the Committee on Research Advisory to the U.S. Department of Agriculture*. Washington, D.C.: National Academy of Sciences.

———. 1989a. *Investing in Research: A Proposal to Strengthen the Agricultural, Food, and Environmental System*. NRC Board on Agriculture. Washington, D.C.: National Academy Press.

———. 1989b. *Alternative Agriculture*. Washington, D.C.: National Academy Press.

———. 1991. *Summary Report 1990: Doctorate Recipients from United States Universities*. NRC Office of Scientific and Engineering Personnel. Washington, D.C.: National Academy Press.

National Science Foundation. 1988. *Characteristics of Doctoral Scientists and Engineers in the United States: 1987*. NSF 88–331, Surveys of Science Resource Series. Detailed statistical tables. Washington, D.C.

———. 1991. *Science and Engineering Doctorates: 1960–90*. NSF 91–310 Final. Surveys of Science Resources Series. Detailed statistical tables. Washington, D.C.

Nelkin, Dorothy. 1984. *Science As Intellectual Property: Who Controls Research?* New York: Macmillan Co.

Nixon, Scott W. 1980. "Between Coastal Marshes and Coastal Waters—A Review of Twenty Years of Speculation and Research on the Role of Salt Marshes in Estuarine Productivity and Water Chemistry." In *Estuarine and Wetland Processes, with Emphasis on Modeling*, edited by Peter Hamilton and Keith B. Macdonald, 437–525. New York: Plenum Press.

Nopar, Doug. 1990. "Participatory Research: Research by and for the People." *Land Stewardship Letter* 8(2): 4. Marine, Minn.: Land Stewardship Project.

Norgaard, Richard B. 1983. "The Scientific Basis of Agroecology." In *Agroecology: The Scientific Basis of Alternative Agriculture*, by Miguel A. Altieri, 7–10. Berkeley: Division of Biological Control, University of California.

Nowak, Peter J. 1984. "Adoption and Diffusion of Soil and Water Conservation Practices." In *Future Agricultural Technology and Resource Conservation*, edited by Burton C. English, James A. Maetzold, Brian R. Holding, and Earl O. Heady, 214–37. Ames: Iowa State Univ. Press.

———. 1987. "The Adoption of Agricultural Conservation Technologies: Economic and Diffusion Explanations." *Rural Sociology* 52(2): 208–20.

Odum, Howard T. 1983. *Systems Ecology: An Introduction*. New York: Wiley Interscience Publications.

Office of Technology Assessment. 1986. *Demographic Trends and the Scientific and Engineering Work Force*. Science Policy Study Background Report no. 9. House Committee on Science and Technology Policy, 99th Cong., 2nd sess. Washington, D.C.: U.S. Government Printing Office.

Paarlberg, Don. 1980. *Farm and Food Policy: Issues of the* 1980s. Lincoln: Univ. of Nebraska Press.

Panel on Alternate Approaches to Graduate Education. 1973. *Scholarship for Society*. Princeton, N.J.: Educational Testing Service.

Papendick, Robert I., Lloyd F. Elliott, and Robert B. Dahlgren. 1986. "Environmental Consequences of Modern Production Agriculture: How Can Alternative Agriculture Address These Concerns?" *American Journal of Alternative Agriculture* 1(1): 3–10.

Paul, Eldor A., and G. Phillip Robertson. 1989. "Ecology and the Agricultural Sciences: A False Dichotomy?" *Ecology* 70(6): 1594–97.

Petersdorf, Robert G. 1986. "Medical Schools and Research: Is the Tail Wagging the Dog?" *Daedalus* 115:99–118.

Pimentel, David, Thomas W. Culliney, Imo W. Buttler, Douglas J. Reinemann, and Kenneth B. Beckman. 1989. "Low-Input Sustainable Agriculture Using Ecological Management Practices." *Agriculture, Ecosystems and Environment* 27:3–24.

Reganold, John P., Lloyd F. Elliott, and Yvonne L. Unger. 1987. "Long-Term Effects of Organic and Conventional Farming on Soil Erosion." *Nature* 330:370–72.

Rhoades, Robert E., Douglas E. Horton, and Robert H. Booth. 1986. "Anthropologist, Biological Scientist and Economist: The Three Musketeers or Three Stooges of Farming Systems Research?" In *Social Sciences and Farming Systems Research: Methodological Perspectives on Agricultural Development*, edited by Jeffrey R. Jones and Ben J. Wallace, 21–40. Boulder, Colo.: Westview Press.

Richards, Stewart A. 1988. "Higher Education for an Alternative Agriculture: British Universities and the Need for Reform." *Biological Agriculture and Horticulture* 5(4): 347–64.

Rosenberg, Charles E. 1964. "The Adams Act: Politics and the Cause of Scientific Research." *Agricultural History* 38(1): 3–12.

———. 1966. "Science and American Social Thought." *Science and Society in the*

United States, edited by David D. Van Tassel and Michael G. Hall, 135–66. Homewood, Ill.: Dorsey Press.

———. 1971. "Science, Technology, and Economic Growth: The Case of the Agricultural Experiment Station Scientist, 1875–1914." *Agricultural History* 45(1): 1–20.

———. 1976. *No Other Gods: On Science and American Social Thought.* Baltimore: Johns Hopkins Univ. Press.

Rossiter, Margaret W. 1979. "The Organization of the Agricultural Sciences." In *The Organization of Knowledge in Modern America, 1860–1920,* edited by Alexandra Oleson and John Voss, 211–48. Baltimore: Johns Hopkins Univ. Press.

Ruttan, Vernon W. 1978. "Reviewing Agricultural Research Programmes." *Agricultural Administration* 5: 1–19.

———. 1991. "Moral Responsibility in Agricultural Research." In *Beyond the Large Farm: Ethics and Research Goals for Agriculture,* edited by Paul B. Thompson and Bill A. Stout, 107–23. Boulder, Colo.: Westview Press.

Rzewnicki, Phil E., Richard Thompson, Gary W. Lesoing, Roger W. Elmore, Charles A. Francis, Anne M. Parkhurst, and Russell S. Moomaw. 1988. "On-Farm Experiment Designs and Implications for Locating Research Sites." *American Journal of Alternative Agriculture* 3(4): 168–73.

Sagoff, Mark. 1985. "Fact and Value in Ecological Science." *Environmental Ethics* 7(2): 99–116.

Sahlins, Marshall. 1976. *The Use and Abuse of Biology: An Anthropological Critique of Sociobiology.* Ann Arbor: Univ. of Michigan Press.

Sailer, R.I. 1979. "Use of Biological Control Agents." In *Introduction to Crop Protection,* edited by W.B. Ennis, Jr., 139–51. Madison, Wis.: American Society of Agronomy and Crop Science Society of America.

Saxberg, Borje O., William T. Newell, and Brian W. Mar. [1981] 1986. "Interdisciplinary Research—A Dilemma for University Central Administration." *Society of Research Administrators Journal* 13 (Fall): 25–43. Reprint (abridged). In *Interdisciplinary Analysis and Research: Theory and Practice of Problem-Focused Research and Development,* edited by Daryl E. Chubin, Alan L. Porter, Frederick A. Rossini, and Terry Connolly, 193–203. Mt. Airy, Md.: Lomond Publications.

Schaller, Neill. 1991. "Background and Status of the Low-Input Sustainable Agriculture Program." In *Sustainable Agriculture Research and Education in the Field,* 22–31. Washington, D.C.: National Academy Press.

Schuh, G. Edward. 1984. "Revitalizing the Land Grant University." Paper presented at colloquium, Strategic Management Research Center, University of Minnesota, Sept. 28.

Schweikhardt, David B., and James T. Bonnen. 1986. "Policy Conflicts in Agricultural Research: Historical Perspective and Today's Challenges." In *The Agricultural Scientific Enterprise: A System in Transition,* edited by Lawrence Busch and William B. Lacy, 13–27. Boulder, Colo.: Westview Press.

Scofield, A.M. 1986. "Organic Farming—The Origin of the Name." *Biological Agriculture and Horticulture* 4(1): 1–5.

Seitz, Wesley D., and Earl R. Swanson. 1980. "Economics of Soil Conservation from the Farmer's Perspective." *American Journal of Agricultural Economics* 62(5): 1084–88.

Shaner, W.W., P.F. Philipp, and W.R. Schmehl. 1982. *Farming Systems Research and Development: Guidelines for Developing Countries.* Boulder, Colo.: Westview Press.

Shapiro, Charles A., William L. Kranz, and Anne M. Parkhurst. 1989. "Comparison of Harvest Techniques for Corn Field Demonstrations." *American Journal of Alternative Agriculture* 4(2): 59–64.

Shennan, Carol, Laurie E. Drinkwater, Ariena H.C. van Bruggen, Deborah K. Letourneau, and Fekede Workneh. 1991. "Comparative Study of Organic and Conventional Tomato Production Systems: An Approach to On-Farm Systems Studies." In *Sustainable Agriculture Research and Education in the Field,* 109–32. Washington, D.C.: National Academy Press.

Simons, Joseph H. 1960. "Scientific Research in the University." *American Scientist* 48:80–90.

Soule, Judith D., and Jon K. Piper. 1992. *Farming in Nature's Image: An Ecological Approach to Agriculture.* Washington, D.C.: Island Press.

Spalding, V.M. 1903. "The Rise and Progress of Ecology." *Science* 17:201–10.

Spencer, Gregory. 1989. *Projections of the Population of the United States, by Age, Sex, and Race: 1988 to 2080.* Current Population Reports, Population Estimates and Projections. Series P-25, no. 1018. Bureau of the Census, U.S. Department of Commerce. Washington, D.C.: U.S. Government Printing Office.

Stansbury, Dale L. 1986. "The Context and Implications of the National Agricultural Research, Extension, and Teaching Act of 1977." In *New Directions for Agriculture and Agricultural Research,* edited by Kenneth A. Dahlberg, 132–46. Totowa, N.J.: Rowman & Allanheld.

Stetten, DeWitt, Jr. 1986. "Publication: Numbers and Quality" (letter). *Science* 232:11.

Stevenson, Steve, and Rick Klemme. 1991. "CIAS Models Public Input for Land Grant Universities." *CIAS Connections* 2(1): 1–3, 8. Center for Integrated Agricultural Systems, Univ. of Wisconsin, Madison.

Stinner, Benjamin R., and Garfield J. House. 1987. "Role of Ecology in Lower-Input, Sustainable Agriculture: An Introduction." *American Journal of Alternative Agriculture* 2(4): 146–47.

———. 1989. "The Search for Sustainable Agroecosystems." *Journal of Soil and Water Conservation* 44(2): 111–16.

Storer, Norman W. 1972. "Relations Among Scientific Disciplines." In *The Social Contexts of Research,* edited by Saad Z. Nagi and Ronald G. Corwin, 229–68. London: Wiley-Interscience.

Storr, Richard J. 1973. *The Beginning of the Future: A Historical Approach to Graduate Education in the Arts and Sciences.* New York: McGraw-Hill Book Co.

Sturtevant, E.L. 1882. "Agricultural Experiments: What the Farmer Wants to Know, and Why He Wants to Know It." In *Fifteenth Annual Report of the Secretary of the Connecticut Board of Agriculture, 1881–82,* pp. 39–51. Hartford: State of Connecticut.

Sumberg, J., and C. Okali. 1989. "Farmers, On-Farm Research and New Technology." In *Farmer First: Farmer Innovation and Agricultural Research,* edited by Robert Chambers, Arnold Pacey, and Lori Ann Thrupp, 109–14. New York: Bootstrap Press.

Sustainable Agriculture Operations Committee. 1991. *Guidelines for Grants Programs on Sustainable Agriculture.* Cooperative State Research Service, U.S. Department of Agriculture. Washington, D.C.

Swanson, Burton E. 1986. "The Contribution of U.S. Universities to Training Foreign Agricultural Scientists." In *The Agricultural Scientific Enterprise: A System in Transition,* edited by Lawrence Busch and William B. Lacy, 229–38. Boulder, Colo.: Westview Press.

Taylor, Carl C. 1941. "Social Science and Social Action in Agriculture." *Social Forces* 20 (December): 154–59.

———. 1947. "Sociology and Common Sense." *American Sociological Review* 12(1): 1–9.

Taylor, Peter J. 1988. "Technocratic Optimism, H.T. Odum, and the Partial Transformation of Ecological Metaphor After World War II." *Journal of the History of Biology* 21(2): 213–44.

Teich, Albert H. [1979] 1986. "Research Centers and Non-Faculty Researchers: Implications of a Growing Role in American Universities." Presented at First International Conference on Interdisciplinary Research Groups, West Germany. Reprint (abridged). In *Interdisciplinary Analysis and Research: Theory and Practice of Problem-Focused Research and Development,* edited by Daryl E. Chubin, Alan L. Porter, Frederick A. Rossini, and Terry Connolly, 215–27. Mt. Airy, Maryland: Lomond Publications.

Thompson, Richard, and Sharon Thompson. 1990. "The On-Farm Research Program of Practical Farmers of Iowa." *American Journal of Alternative Agriculture* 5(4): 163–67.

Thornley, J.H.M., and C.J. Doyle. 1984. "The Management of Publicly-Funded Agricultural Research and Development in the U.K.: Questions of Autonomy and Accountability." *Agricultural Systems* 15:195–208.

Thornley, Kay. 1990. "Involving Farmers in Agricultural Research: A Farmer's Perspective." *American Journal of Alternative Agriculture* 5(4): 174–77.

Tivy, Joy. 1990. *Agricultural Ecology.* Harlow, Essex, England: Longman Scientific & Technical (copublished with John Wiley & Sons, New York).

U.S. Department of Agriculture. 1980. *Report and Recommendations on Organic Farming.* USDA Study Team on Organic Farming. Washington, D.C.

———. 1992. *National Research Initiative Competitive Grants Program. Program Description: Guidelines for Proposal Preparation and Submission.* USDA Cooperative State Research Service. Washington, D.C.

van den Berghe, Pierre L. 1978. "Bridging the Paradigms: Biology and the Social Sciences." In *Sociobiology and Human Nature,* edited by Michael S. Gregory, Anita Silvers, and Diane Sutch, 33–52. San Francisco: Jossey-Bass Publishers.

Vollmer, Howard M. 1972. "Basic and Applied Research." In *The Social Contexts of Research,* edited by Saad Z. Nagi and Ronald G. Corwin, 67–96. London: Wiley-Interscience.

Wagstaff, Howard. 1987. "Husbandry Methods and Farm Systems in Industrialised Countries Which Use Lower Levels of External Inputs: A Review." *Agriculture, Ecosystems and Environment* 19:1–27.

Wallace, Henry. 1936. "Report of the Secretary of Agriculture to the President of the United States." In *Yearbook of Agriculture, 1936,* pp. 1–117. U.S. Department of Agriculture. Washington, D.C.

Walters, D.T., D.A. Mortensen, C.A. Francis, R.W. Elmore, and J.W. King. 1990. "Specificity: The Context of Research for Sustainability." *Journal of Soil and Water Conservation* 45(1): 55–57.

Watkins, Gordon. 1990. "Participatory Research: A Farmer's Perspective." *American Journal of Alternative Agriculture* 5(4): 161–62.

Webster, F.W. 1914. "Bringing Applied Entomology to the Farmer." In *Yearbook of the Department of Agriculture, 1913*, pp. 75–92. Washington, D.C.: U.S. Government Printing Office.

Weinberg, Alvin M. 1967. *Reflections on Big Science*. Cambridge, Mass.: M.I.T. Press.

———. 1970. "Scientific Teams and Scientific Laboratories." *Daedalus* 99(4): 1056–75.

Weiss, Paul. 1964. "Science in the University." *Daedalus* 93: 1184–1218.

Wernick, Sarah, and William Lockeretz. 1977. "Motivations and Practices of Organic Farmers." *Compost Science* 18(6): 20–24.

Wiley, H.W. 1904. "Soil Fertility." In *Proceedings of the Seventeenth Annual Convention of the Association of American Agricultural Colleges and Experiment Stations*, 142–46. Bulletin 142. Office of Experiment Stations, U.S. Department of Agriculture. Washington, D.C.

Wilson, Edward Osborne. 1975. *Sociobiology: The New Synthesis*. Cambridge, Mass.: Belknap Press, Harvard Univ. Press.

Winrock International Conference Center. 1982. *Science for Agriculture. Report of a Workshop on Critical Issues in American Agricultural Research*. New York: Rockefeller Foundation.

Wischmeier, W. H. 1976. "Use and Misuse of the Universal Soil Loss Equation." *Journal of Soil and Water Conservation* 31(1): 5–9.

———. 1984. "The USLE: Some Reflections" (interview). *Journal of Soil and Water Conservation* 39(2): 105–7.

Wischmeier, W.H., and D. D. Smith. 1978. *Predicting Rainfall Erosion Losses—A Guide to Conservation Planning*. Agricultural Handbook no. 537. Science and Education Administration, U.S. Department of Agriculture. Washington, D.C.

Worster, Donald. 1985. *Nature's Economy: A History of Ecological Ideas*. Cambridge: Cambridge Univ. Press.

———. 1991. "The Marriage of Ecology and Agriculture." *Journal of Production Agriculture* 4(3): 279–82.

Index

Academic freedom, 125

Adams Act (1906), 24, 119

Advisory committees: in alternative agriculture research, 22; as alternatives to farmer-controlled research, 125; farmers as members of, 117; and grant guidelines, 172; makeup of, 181–83; in recent legislation, 22; under Research and Marketing Act, 182; researchers as members of, 189

Advocacy groups, 12, 100, 115, 125, 172. *See also* Interest groups

Affirmative action, 195

Agrarianism, 121, 124

Agricultural chemicals: information needed to use, 78, 91, 94, 98; and site-specificity, 103; ways to reduce use of, 5

Agricultural chemistry, effect on U.S. agriculture, 91

Agricultural colleges: contrasting orientations of, 141; declining support of, 191; departmental structure of, 132–33, 138–40; ecology at, 77; isolation of, 198; non-departmental units within, 139–40; possibilities for change in, 206

Agricultural development: agroecology as strategy for, 73; education for, 191–92

Agricultural disciplines, 64; communication among, 186–87; competition among, 134–36; distorting effects of, 156; emergence of, 146; functions of, 26, 61; and professional rewards, 150; proliferation of, 62; relation to agricultural problems, 49; role in quality control, 143; synthesis of, 52, 57–60, 62

Agricultural economics: applied nature of, 146; origin of, 63; relation to general economics, 63–64; relation to other disciplines, 135, 137; scope of, 132

Agricultural education: accommodating non-traditional students in, 196; backgrounds of students in, 190; early years of, 189, 202; foreign students in, 191; goals of, 175–76, 185–87, 199; importance of diverse faculty in, 196–97; in-

237